E. Vesentini (Ed.)

Geometry of Homogeneous Bounded Domains

Lectures given at a Summer School of the
Centro Internazionale Matematico Estivo (C.I.M.E.),
held in Urbino (Pesaro), Italy,
July 3-13, 1967

Springer

FONDAZIONE
CIME
ROBERTO CONTI

C.I.M.E. Foundation
c/o Dipartimento di Matematica "U. Dini"
Viale Morgagni n. 67/a
50134 Firenze
Italy
cime@math.unifi.it

ISBN 978-3-642-11059-7 e-ISBN: 978-3-642-11060-3
DOI:10.1007/978-3-642-11060-3
Springer Heidelberg Dordrecht London New York

Reprint of the 1st Ed. C.I.M.E., Ed. Cremonese, Roma 1968
With kind permission of C.I.M.E.

Printed on acid-free paper

Springer.com

CENTRO INTERNAZIONALE MATEMATICO ESTIVO

(C.I.M.E.)

3° Ciclo - Urbino 5-13 Luglio 1967

GEOMETRY OF HOMOGENEOUS BOUNDED DOMAINS

Coordinatore : Edoardo Vesentini

GINDIKIN, S. G., PJATECCKIĬ-ŠAPIRO I. I., VINBERG E. B. : Homogeneous Kähler manifolds pag. 1

GREENFIELD S. J. : Extendibility properties of real Submanifolds of $\mathbb{C}^n$. pag. 89

KAUP W. : Holomorphie Abbildungen in Hyperbolische Räume. pag. 109

KORANYI A. : Holomorphic and harmonic functions on bounded symmetric domains. pag. 125

KOSZUL J. L. : Formes harmoniques vectorielles sur les espaces localement symétriques pag. 197

MURAKAMI S. : Plongements holomorphes de domaines symetriques. pag. 261

STEIN E. M. : The analogues of Fatous's theorem and estimates for maximal functions pag. 287

CENTRO INTERNAZIONALE MATEMATICO ESTIVO
(C.I.M.E.)

S.G. GINDIKIN, I.I. PJATECCKIĬ-ŠAPIRO, E.B. VINBERG

"HOMOGENEOUS KÄHLER MANIFOLDS"

Corso tenuto ad Urbino dal 5 al 13 luglio 1967

HOMOGENEOUS KÄHLER MANIFOLDS [1]

by

S. G. GINDIKIN, I. I. PJATECCKIĬ-ŠAPIRO, E. B. VINBERG

Introduction. Recall of certain results.

1. Definition of homogeneous Kähler manifolds.

Let $h = g + i$ be a positive definite Hermitian differential form on the complex manifold M. Then g is a positive definite symmetric differential form and η is a non-degenerate skew-symmetric differential form of type (1,1), and

$$g(x, y) = \eta(Ix, y) \tag{1}$$

where I is the complex structure operator. The complex manifold M with the positive definite hermitian differential form h is called Kählerian if one of the following equivalent conditions is satisfied

(K1) $$d = 0 ;$$

(K2) The parallel translation with respect to the riemannian metric g preserves the complex structure of the tangent space, i.e. $\nabla I = 0$;

(K3) In local coordinates z^{α}, $\bar{z}^{\alpha}$ the coefficients $h_{\alpha\bar{\beta}}$ of the form h can be represented in the form

$$h_{\alpha\bar{\beta}} = \frac{\partial^2 \log \varphi}{\partial z^{\alpha} \partial \bar{z}^{\beta}} \tag{2}$$

where φ is a positive real function.

The prof of the equivalence of conditions (K1) - (K3) can be found for example in [13, 27].

An automorphism of the Kähler manifold M is an invertible

---

[1] English Translation by Adam Koranyi.

holomorphic map preserving the form h. We shall denote the group of all automorphisms of the Kähler manifold M by $G(M)$; its connected component by $G^o(M)$. We shall also consider the group $G_A(M)$ of all invertible holomorphic transformations of the manifold M and the group $G_R(M)$ of all isometries of M as a riemannian manifold. Then

$$G(M) = G_A(M) \cap G_R(M) .$$

We denote by $G_A^o(M)$ and $G_R^o(M)$ the connected components of the groups $G_A(M)$ and $G_R(M)$ respectively. In [8, 12] some sufficient conditions are given in order that $G_R^o(M) = G^o(M)$. We are not going to discuss these conditions here. However the connection between the groups $G_A(M)$ and $G(M)$ will be considered in certain cases.

The Kähler manifold M is called homogeneous if the group $G(M)$ acts transitively on it.

Often the homogeneity of a Kähler manifold is defined by the transitivity of the group $G_A(M)$. From the results of A. Borel - R. Remmert [3] and of Tits [22] it follows that, if a compact Kähler manifold is homogeneous in this sense, then there exists on it a Kählerian structure (compatible with the given complex structure) with respect to which it is a homogeneous in our sense.

In the non-compact case, it is unlikely that the consideration of homogeneous complex manifolds carrying Kählerian structures will lead to a significative classification.

The simplest examples of homogeneous Kähler manifolds are the hermitian space H^n, the complex torus T^n, the complex projective space P^n and the unit disc K in the complex plane. In the following three paragraphs we shall describe three fundamental types of homogeneous Kähler manifolds which have an extremely important significance for the theory.

In the following we shall abbreviate the words "homogeneous Kähler manifolds" by "h.K.m.".

2. Locally flat homogeneous Kähler manifolds.

These are h.K.m.'s which have zero curvature in the Riemannian metric g. They are easy to classify. First of all, every homogeneous locally flat h.K.m. is isomorphic with the hermitian space H^n In fact, by a known theorem of E. Cartan it is isomorphic to an Euclidean space as a riemannian manifold; from the Kähler condition (K2) it follows that the complex structure is invariant under parallel translations.

Any locally flat h.K.m. can be obtained by factoring H^n by some lattice.

The group $G^o_A(M)$ for a locally flat h.K.m. M is the complex hull of the groups $G^o(M)$. A maximal complex subgroup of $G^o(M)$ is the group of parallel translations. It is transitive on M.

3. Simply connected compact homogeneous Kähler manifolds.

These h.K.m.'s have been studied by several authors and have been completely classified (Lichnerowicz [11] , Borel [2] , Wang [26]). We note that Wang [26] found all simply connected complex homogeneous manifolds. Some of these do not admit any Kählerian structure.

We formulate the fundamental result concerning this type of h.K.m. . Let M be a simply connected compact h.K.m. . Then the group $G^o(M)$ is a compact semi-simple Lie group with trivial center, its isotropy subgroup is connected and is the centralizer of a torus. Conversely, if G is a connected compact Lie group and K is the centralizer of a torus in G , then there exists an invariant Kähler structure on the homogeneous space G/K . Every complex Lie group has only finitely many sub-groups (up to conjugation) that are centralizers of tori.

They can all be easily found.

Every simply connected compact h.K.m. M can be realized as an algebraic manifold in a complex projective space P^n in such a way that the automorphisms of M will be the restrictions of unitary projective transformations. However the Kähler structure of M will in general be different from the Kähler structure induced by P^n.

As in the locally flat case, the group $G^o_A(M)$ of the simply connected compact h.K.m. M is the complex hull of $G^o(M)$. However in this case $G^o(M)$ has no non-trivial complex subgroups. The isotropy subgroup K_A of the group $G^o_A(M)$ is connected and contains a Borel subgroup (the group K_A is <u>not</u> the complex hull of the group K).

Let us look at a typical example.

Let G be the group of $n \times n$ unitary matrices. Let $K = K(n_1, \ldots, n_s)$, $\sum n_i = n$ be the subgroups consisting of all diagonal block matrices of order $n_1, \ldots, n_s$. We call an $(n_1, \ldots, n_s)$ flag a sequence of subspaces of the hermitian space H^n of dimensions $n, n_1 + n_2, \ldots, n_s + \ldots + n_{s-1}$, contained successively in each other. The homogeneous manifold G/K can be realized as the manifold of $(n_1, \ldots, n_s)$ flags.

In a natural way it is contained in a complex projective space ; the Kähler structure induced by this inclusion is invariant under the group G .

We should mention that the group G in this example acts non effectively on $G/_K$. The kernel of the action is the center of the group G, which is contained in K . This is in agreement with the general theory, since the automorphisms group of a simply connected compact h.K.m. always has a trivial center as we remarked before.

The group G_A of all non-singular complex $n \times n$ matrices is the complex hull of the group G and acts analytically on $G/_K$, but it does not preserve the Kähler structure. The isotropy subgroup of G_A is the the group $K_A = K_A(n_1, \ldots, n_s)$ which consists of all triangular block matrices with blocks of order $n_1, \ldots, n_s$ on the diagonal.

This is how one describes (up to the choice of the Kählerian structure) all simply connected complex h.K.m. which are connected with the unitary group. For the other compact Lie groups there is an analogous construction.

Matsushima [14] proved that every compact h.K.m. is a direct product of a simply connected compact h.K.m. and a complex torus.

4. Homogeneous bounded domains.

Let D be a bounded domain in the n-dimensional complex space $\mathbb{C}^n$. The Bergman metric [1, 5, 27] defines in D a canonical Kählerian structure. This structure is invariant under all analytic automorphisms of the domain D, that is, now $G_A(D) = G(D)$.

The domain D is said to be homogeneous if the group $G_A(D)$ is transitive on it.

In the case of a homogeneous domains the coefficients of the Bergman metric can be found on the basis of (2) where for φ one has to take the density of the invariant measure. Beside the canonical Kählerian structure there may exist other Kählerian structures in a homogeneous domain which are invariant with respect to $G_A(D)$, or with respect to some transitive subgroup of it. Differently from the case of the other fundamental types of h.K.m., for the bounded domains D, the group $G_A(D)$ does not contains any non-trivials complex subgroup.

In the following we shall abbreviate the word "homogeneous bounded domains" as "h.b.d.".

In $\mathbb{C}^1$, up to analytic isomorphisms, there is only one h.b.d.: the unit disc $\{|z| < 1\}$. In $\mathbb{C}^2$ there exist two non-isomorphic h.b.d.'s: the complex ball

$$\left\{ |z_1|^2 + |z_2|^2 < 1 \right\}$$

and the bi-cylinder

$$\left\{ |z_1| < 1 , |z_2| < 1 \right\}.$$

The non-isomorphy of these domains was proved by Poincaré [15]. The non-existence of other h.b.d. in $\mathbb{C}^2$ was shown by E. Cartan [4]. He also found all h.b.d.'s in $\mathbb{C}^3$.

The domain $D \subset \mathbb{C}^4$ is called <u>symmetric</u> if for every point $z \in D$ there exists an involutive analytic automorphism b_z of D for which z is a unique fixed point.

Every symmetric bounded domain is homogeneous and is a symmetric space.

Using the classification of symmetric spaces, E. Cartan enumerated all bounded symmetric domains [4]. In the same work he established that for $n \leqslant 3$ all h.b.d.'s are symmetric. In connection with this he posed the problem : are all h.b.d.'s symmetric? And if not, how can one construct them ?

A. Borel [2] and Koszul [9] showed that if a h.b.d. there is acted upon by semi-simple group of analytic automorphisms , then their domain is symmetric. The same result, with still weaker hypotheses, was proved by Hano [7].

In [16] Pjatecckii - Šapiro obtained a negative answer to the first part of E. Cartan's problem. He constructed an example of a non-symmetric h.b.d. in $\mathbb{C}^4$. (We shall describe it in § 6.)

It turned out later that the symmetric domains are in a certain sense exceptional among the h.b.d.'s in $\mathbb{C}^n$, while for every n there are only finitely many bounded symmetric domains in $\mathbb{C}^n$. It is interesting that the non-symmetric h.b.d.'s arise naturally in connection with the study of homogeneous fiberings of symmetric domains [17, 21] .

In [18, 19, 20] Pjatecckii-Šapiro studied in detail those h.b.d.'s which admit a transitive solvable group of automorphisms acting without fixed points. He proved that every such h.b.d. is isomorphic with a non-bounded homogeneous domain, which is homogeneous under a group of affine transformations (a description of these domains, so-called Siegel domains of type I and II, will be given in the first part of these lectures) ;

In the joint work [25] by Vinberg, Gindikin and Pjatecckii-Šapiro the same result was obtained without any restrictive hypothesis. It turned out a posteriori that the condition imposed by Pjatecckii-Šapiro is not really a restriction. For every h.b.d. D (with any homogeneous Kählerian structure) in the group $G^{o}(D)$ there is a transitive splittable solvable subgroup T(D)) acting on D without fixed points. There exists a realization of D as a convex unbounded domain such that the elements of the group T(D) are affine transformations. The group $G^{o}(D)$ has no center. The isotropy subgroup is a maximal compact subgroup in $G^{o}(D)$.

All these results were obtained in [25] . In these lectures we prove some theorems about h.K.m.'s from which the result of Pjatecckii-Šapiro follows under the hypothesis that the domain D admits a transitive splittable solvable group of automorphisms.

5. The structure of arbitrary homogeneous Kähler manifolds.

Every h.K.m. which has a transitive semi-simple group of automorphisms admits a holomorphic fibering with a simply connected h.K.m. as its fiber, the base of which is analitycally isomorphic with a symmetric bounded domain (Borel [2], Matsuhima [14]). Every h.K.m. which has a transitive reductive group of automorphisms decomposes into the direct product of an h.K.m. admitting a transitive semi-simple group of automorphisms and of a locally flat h.K.m. (Matsushima [14]).

These theorems and several other results, some of which will be discussed below, gave us the basis for the following conjecture.

Fundamental conjecture. Every homogeneous Kähler manifold admits a holomorphic fibering, the base of which is analytically isomorphic with a homogeneous bounded domain, and the fiber, with the induced Kähler structure, is isomorphic with the direct product of a locally flat h.K.m. and a simply connected compact h.K.m. .

Besides the cases mentioned above (results of A. Borel and Matsushima) this conjecture is essentially proved, even though this is not explicitely mentioned, in our article [25] for h.K.m.'s, which admit a transitive group of automorphisms on which the pre-image of the differential form η = Im h (cf. § 1) is the differential of some left-invariant form. In this case there is no locally flat factor in the fiber.

A considerable part of these lectures will deal with the proof of the fundamental conjecture for Kähler manifolds which admit a transitive splittable solvable group of automorphisms.

This result is due to Vinberg and Gindikin.

Let us make some remarks in connection with the fundamental conjecture. The fibering about which we have spoken is unique, since its fibers can by characterized as the maximal sets on which all bounded holomorphic functions are constant. Therefore it is preserved by all analytic automorphisms of the manifold. Furthermore the base of the fibering, being a h.b.d., is homeomorphic with an affine space. Consequently this fibering is topologically trivial. Its structure group is the group of invertible holomorphic map of the fiber, and is a complex Lie group (cf. §§ 2 and 3). According to a theorem of Grauert [6] such holomorphic fiberings are trivial.

Therefore if the fundamental conjecture is true, then every h.K.m. is, as a complex manifold, isomorphic with the direct product of h.K.m's of the three fundamental types described in § § 2 - 4 .

PART I - Siegel domains .

1. Siegel domains of type I .

We have already spoken in the introduction about the important role played in the theory of h.b.d.'s by their affine homogeneous realizations. In the case of symmetric domains we usually consider their realization as "disc".

In these realizations the isotropy group of some point of the domain consists of linear transformations. Here we shall consider other realizations of the type of the "upper half plane" in which there is a transitive group of affine transformations (this group can be interpreted as the isotropy group of a point of the boundary of the domain). In the course of this, we shall consider certain special classes of affine homogeneous domains : the Siegel domains of type I and II . In this paragraph we shall talk about the following simplest generalization of the upper half plane to the case of several complex variables.

Let V be an open convex cone in the n dimensional real space $\mathbb{R}^n$ (i.e. if $x, y \in V$, then $\lambda x + \mu y \in V$ for $\lambda \geqslant 0$, $\mu \geqslant 0$, $\lambda + \mu \neq 0$) not containing any straightline .

The domain in $\mathbb{C}^n$

(1) $$D(V) = \mathbb{R}^n + i\ V$$

is called a Siegel domain of type I associated with the

cone $V^{(1)}$.

Proposition I. Every Siegel domain of type I is isomorphic with a bounded domain.

Proof. The convex cone V, containing no straight line, is contained in some n-sided angle. Making a linear transformation of $\mathbb{R}^n$, this angle can be transformed into the positive octant of $\mathbb{R}^n$, V' : $y_k > 0$ $(k = 1, \dots, n)$. If this transformation is continued by the same formulas to $\mathbb{C}^n$, then the domain $D(V)$ becomes a part of the domain $D(V')$. The domain $D(V')$, being a direct product of upper half planes, is analytically isomorphic with the n-dimensional circular poly-cylinder

$$\left\{ |z_k| < 1, \ k = 1, \dots, n \right\} .$$

That is the domain $D(V)$ can be mapped into a subset of this poly-cylinder.

We introduce an auxiliary notion. We call skeleton of the domain $D \subset \mathbb{C}^n$ a set Ω_D such that :

α) every function $f(z)$ which is holomorphic on $\bar{D}$ (the closure of D) and assuming its maximum modulus in $\bar{D}$, reaches its maximum modulus in some point of Ω_D ;

β) for every point $z_0 \in \Omega_D$ there exists a function holomorphic in $\bar{D}$ whose modulus assumes its maximum in the point z_0 and only there.

It is clear that the skeleton Ω_D is uniquely defined, if it exists, and that it is preserved by all automorphisms of the domain D which are holomorphic on $\bar{D}$.

Lemma 1. The skeleton of the Siegel domain of type I $D(V)$ is the set $\mathbb{R}^n$

---

[1] For Siegel domain of type I a more widespread name is "radiated tube domains".

Proof: α). Let the maximum modulus of the function $f(z)$ holomorphic in $\bar{D}$ be assumed at the point $z_0 \in D$, $\operatorname{Im} z_0 = 0$.

We may assume that $\operatorname{Re} z_0 = 0$. The function of one variable

$$\varphi(\lambda) = f(\lambda z_0)$$

will be holomorphic in the half plane $\{\operatorname{Re} \lambda > 0\}$ and its modulus reaches its maximum for $\lambda = 1$. But then it assumes its maximum also for $\lambda = 0$.

β) From the considerations of the proof of proposition 1, it is clear that, without restriction of generality, we may assume that the domain $D(V)$ is contained in a direct product of upper half planes

$$\left\{\operatorname{Im} z_k > 0,\ k = 1, \dots, n\right\} .$$

Then for the point $x^0 = (x_1^0, \dots, x_n^0) \in \mathbb{R}^n$, the function

$$f(z) = \frac{1}{(z_1 - x_1^0 + i) \dots (z_n - x_n^0 + i)}$$

will satisfy the condition β) in the definition of the skeleton.

(The point x_0 will be the unique maximum of $|f(z)|$ in the domain $\left\{\operatorname{Im} z_k \geqslant 0,\ k = 1, \dots, n\right\}$ and therefore also in $D(V)$).

Let V be a cone in $\mathbb{R}^n$ having the properties mentioned above.

We denote by $G(V)$ the group of non singular linear transformations of $\mathbb{R}^n$ which preserve V. A cone V is called <u>homogeneous</u> if $G(V)$ acts transitively on V. Analogously to Siegel domain of type I $D \subset \mathbb{C}^n$, we denote by $G_a(D)$ the group of affine transformations of $\mathbb{C}^n$ which preserve D. If the group $G_a(D)$ acts transitively on D, then we shall call D a <u>homogeneous Siegel domain of type</u> I. Here we shall not explain more precisely the term "affine homogeneous Siegel domain" since in the case of Siegel domains one always talks about homogeneity with respect to an affine group.

Proposition 2. The Siegel domain of type I $D(V)$ is homogeneous if, and only if, the cone V is homogeneous.

Proof 1. If V is a homogeneous cone and $G(V)$ is a transitive group of automorphisms on it, then the maps of $\mathbb{C}^n$ of the form

$$(2) \qquad z \longrightarrow A \; z + a \;,$$

where $A \in G(V)$ (more exactly, we have to take the complex continuation of the linear transformations of $\mathbb{R}^n$), $a \in \mathbb{R}^n$, form a transitive group in $D(V)$.

2. We show the converse. Let $D(V) = D$ be a homogeneous Siegel domain of type I, and let (2) be an affine transformation preserving $D(V)$ (A is a non-singular complex linear transformation, $a \in \mathbb{C}^n$). Under such a map the skeleton Ω_D must be preserved. This follows from the general fact that the skeleton is preserved by maps which are analytic on the closure of the domains. However in our case it is also enough to mention that the skeleton Ω_D is a maximal flat component of the boundary of D containing 0. Since the skeleton $\Omega_D = \mathbb{R}^n$ is preserved, the linear transformation A and the vector a must be real. Then for our automorphisms, $y = \mathrm{Im}\, z$ is acted upon by the transformation A and the cone V must be preserved. From this it follows that the linear transformations A entering in (2) form a transitive group of linear automorphisms of the cone V. In order to construct examples of homogeneous Siegel domains of type I, it is sufficient to construct examples of homogeneous convex cones not containing straight lines.

Example 1. Consider the cone V of symmetric positive definite matrices g of order ℓ. This is a convex cone, containing no straight lines, in the space $\mathbb{R}^n$, $n = \frac{\ell(\ell+1)}{2}$, of symmetric matrices of order ℓ . The automorphisms of the cone are the mappings

(3) $$y \longrightarrow g\, y\, g' \quad ,$$

where g is a non-singular matrix of order ℓ , and g' is its transposed. This is the formula for the change of the matrix of a quadratic form under a change of variables. The transitivity of the group $G(V)$ follows form the possibility of reducing every positive definite quadratic form to a sum of squares. We mention that the transitivity is preserved if we restrict ourselves in (3) to triangular (for example upper triangular) matrices with positive diagonal elements .

The corresponding Siegel domain of type I $D(V)$ (in this case it is usually called "Siegel upper half plane") consists of the complex symmetric matrices of order ℓ with positive definite imaginary part.

Example 2. As another example we consider the cone V of complex hermitian positive definite matrices of order ℓ , considered as a cone in the real space $\mathbb{R}^n$, $n = \ell^2$ of hermitian matrices. In it there acts transitively the group of non-singular complex matrices g :

(4) $$y \longrightarrow g\, y\, g^* \quad ,$$

where $g^* = \bar{g}'$; here the group of triangular matrices with real positive diagonal elements acts on V transitively without fixed points.

The corresponding Siegel domain of type I can be realized as the set of these complex matrices z of order ℓ , for which the hermitian matrix $\frac{1}{2i}\,(z - z^*)$ is positive definite.

The Siegel domains of example 1, 2, are symmetric. In both cases, the symmetry at the point $z = i\, E$ (E is the identity matrix) is given by :

(5) $$z \longrightarrow -\,z^{-1}$$

It is clear that it is sufficient to give the involution at one point.

It turns out that a Siegel domain of type I is symmetric if, and only if, the cone V is self-adjoint with respect to some scalar product (The adjoint cone V^* consists of these x for which the inner product $< x , y > > 0$ for all $y \in \bar{V}$, $y \neq 0$). We are not going to prove this result here.

Example 3. In order to construct non-symmetric homogeneous Siegel domains of type I, one has to construct homogeneous non-self-adjoint cones (with respect to any scalar product). Such cones appear first in $\mathbb{R}^5$. Consider the cone in $\mathbb{R}^5$:

$$(6) \qquad \begin{cases} y_{11} \; y_{33} - y_{13}^2 > 0 \\ y_{22} \; y_{33} - y_{23}^2 > 0 \\ y_{33} > 0 . \end{cases}$$

Its adjoint cone is the cone of symmetric positive definite matrices of the form

$$\begin{pmatrix} y_{11} & y_{12} & y_{13} \\ y_{12} & y_{22} & 0 \\ y_{13} & 0 & y_{33} \end{pmatrix} .$$

This cone is not linearly equivalent with the cone (6).

Correspondingly there exist homogeneous non-symmetric Siegel domains of type I in $\mathbb{C}^n$ for $n \geq 5$. Let us recall (cf. also the following paragraph) that non-symmetric h.b.d.'s exist in $\mathbb{C}^n$ for $n \geq 4$. Let us mention also that there exist an analytic continuum of non-isomorphic Siegel domains of type I in $\mathbb{C}^n$, for $n \geq 11$ (in the class of all h.b.d.'s there is a continuum of non isomorphic ones for $n \geq 7$).

2. Siegel domains of type II.

From the concluding remarks of the previous paragraph one can infer that not all h.b.d.'s are isomorphic with Siegel domains of type I. One can get to the same conclusion from simplex considerations too. For the complex ball

$$|z_1|^2 + \ldots + |z_n|^2 < 1$$

with $n \geq 2$, there exists no realization as a Siegel domain of type I. This fact will in full be a consequence of the results of the following part, but let us show right now that the complex ball cannot be mapped onto a Siegel domain of type I by a mapping which is holomorphic on the closed ball. For the proof it is sufficient to remark that the skeleton of a Siegel domain of type I has real dimension $n = \dim_{\mathbb{C}} D$, while the skeleton of the ball coincides with its boundary, i.e. has real dimension $2n-1$. This follows from the fact that the modulus of the function $\frac{1}{z_1 - \frac{1}{2}}$ reaches its maximum only at the point $(1, 0, \ldots, 0)$ in the closed ball, and the group of unitary linear transformations acts transitively on the boundary of the ball. The ball can be mapped onto an affine homogeneous domain by setting

$$z_1 = \frac{z-i}{z+i}, \quad z_2 = \frac{u_1\sqrt{2}}{z+i}, \ldots, \quad z_n = \frac{u_{n-1}\sqrt{2}}{z+i} \quad .$$

We obtain as image the domain

$$\text{(7)} \qquad \operatorname{Im} z - |u_1|^2 - \ldots - |u_{n-1}|^2 > 0 \ .$$

We describe now a transitive group of affine transformations of the domain (7). We consider the maps

$$\text{(8)} \qquad \begin{aligned} z &\to z + a + 2i \sum u_k \bar{c}_k + i \sum c_k{}^2 , \\ u &\to u + c \end{aligned}$$

where $a \in \mathbb{R}$, $c \in \mathbb{C}^{n-1}$

It easy to check that the domain (7) is preserved by the mappings (8) . Besides we have the automorphisms

$$(9) \qquad \begin{array}{l} z \to \lambda^2 z , \\ u \to \lambda u \end{array} \qquad (\lambda > 0) .$$

The mappings (8) , (9) generate a transitive group. In fact any point (z, u) satisfying (7) can be mapped by (8) onto a point $(iy, 0)$ $(y > 0)$, and this point can be mapped by (9) onto $(i, 0)$. For the proof of the transitivity of the group of automorphisms it is enough to prove that an arbitrary point of the domain can be carried onto some given point.

The construction (7) admits the following generalization.

Let V be a convex cone in R^n not containing straight lines. The map $F : \mathbb{C}^m \times \mathbb{C}^m \to \mathbb{C}^n$ will be called a V - <u>hermitian form</u> if

(10) $F(\lambda_1 u_1 + \lambda_2 u_2, v) = \lambda_1 F(u_1, v) + \lambda_2 F(u_2, v) \quad (\lambda_1, \lambda_2 \in \mathbb{C}; u_1, u_2, v \in \mathbb{C}^m)$,

(11) $F(u, v) = \overline{F(v, u)}$,

(12) $F(u, u) \in \bar{V}$, where $\bar{V}$ is the closure of the cone V ,

(13) $F(u, u) = 0$ only if $u = 0$.

In the case where V is the positive half line, a V -hermitian form is a usual positive definite hermitian form .

<u>Siegel domain of type II</u> $D(V, F)$ associated to the cone V and to the V-hermitian form F is the domain in $\mathbb{C}^{n+m}$ consisting of the points (z, u), $z \in \mathbb{C}^n$, $u \in \mathbb{C}^m$ satisfying the condition

$$(14) \qquad \operatorname{Im} z - F(u, u) \in V .$$

For $n = 1$ we obtain the domain (7) . The Siegel domains of type I can be considered as special cases of Siegel domains of type II $(m = 0)$.

First of all we prove following

Proposition 3 . Every Siegel domain of type II, $D(V,F)$, is analytically isomorphic will some bounded domain .

Proof. Since the V-hermitian form F becomes a V'-hermitian form if we change the cone V to a cone $V' \supset V$, we can include the domain $D(V,F)$ in a domain $D(V',F)$, where V' is an octant. So we can restrict ourselves to the case where V is such an octant, and we can assume that it is the positive octant of $\mathbb{R}^n$ (by making a linear transformation if necessary). In this case all components of F : F_1, ..., F_n will be non-negative definite hermitian forms.

We represent each of the forms F_k as a sum of squares of moduli of linear forms

$$F_k(u,u) = \sum_j \left| L_{jk}(u) \right|^2 . \tag{15}$$

From the set of all forms L_{jk} we choose a maximal set S of linearly independent forms is equal to m since by (13) the forms L_{jk} have a unique common zero. In the sums (15) we replace by 0 all forms which do not occur in the system S .

We denote the resulting hermitian forms by $\tilde{F}_k$. The domain $D(V,\tilde{F})$ contains the domain $D(V,F)$. Choosing the forms in S as new variables in $\mathbb{C}^m$ we obtain that the domain $D(V,F)$ in these variables is the direct product of n domains of the form (7) , that is analytically isomorphic with the direct product of n balls.

The proof is complete.

Now we consider the question of the automorphisms of a Siegel domain of type II. A Siegel domain of type II $D \subset \mathbb{C}^{n+m}$ is called homogeneous if the group $G_a(D)$ of these affine transformations of $\mathbb{C}^{n+m}$ which preserve D acts transitively on D In a Siegel domain of type II we have always the analogues of the transformations (8) :

$$(16)\qquad \begin{cases} z \to z + a + 2i\ F(u,c) + i\,(F(c,c) , \\ u \to u + c \qquad\qquad (a \in \mathbb{R}^n ,\ c \in \mathbb{C}^m) . \end{cases}$$

Before determining the general form of affine automorphisms , we study the skeleton of Siegel domains of type II .

Lemma 2. The skeleton Ω_D of the Siegel domain of type II $D = D(V,F)$ consists of those points (z,u) for which $\operatorname{Im} z = F(u,u)$.

Proof. Let $f(z)$ be a holomorphic function in $\bar{D}$ whose modulus assumes its maximum at the point $(z_0, u_0) \in \bar{D}$, $\operatorname{Im} z_0 \neq F(u_0, u_0)$. Using the mappings (16) , we can assume that $u_0 = 0$, $\operatorname{Re} z_0 = 0$, i.e. $z_0 = i\,y_0$, $y_0 \in \bar{V}$, $y_0 \neq 0$. Then , just as in the proof of Lemma 1, the function $\varphi(\lambda) = f(\lambda z_0, 0)$ will be holomorphic in the upper half-plane, and its modulus assume its maximum in the point $\lambda = 1$.

Using the mapping (16) it is enough to prove property (β) for $u_0 = 0$. In that case, if $(z_0, 0) \in \overline{D(V,F)}$, then $z_0 \in \overline{D(V)}$, and the function constructed in the proof of lemma 1 will satisfy the required conditions (it is essential that when $(z,u) \in \overline{D(V,F)}$, then $z \in \overline{D(V)}$, and $\operatorname{Im} z \neq 0$ for $u \neq 0$ by (13)) .

Now we shall study the general form of the automorphisms if a Siegel domain of type II.

Proposition 4. Every affine automorphism preserving the Siegel domain $D(V,F)$ of type II is of the form :

$$(17)\qquad \begin{cases} z \to A\ z + a + 2i\ F(B\ u, c) + i\ F(c,c) \\ u \to B\ u + c \end{cases}$$

where $a \in \mathbb{R}^n$, $c \in \mathbb{C}^m$, A is an automorphism of the cone V, B is a linear transformation of $\mathbb{C}^m$ such that

$$(18)\qquad AF(u,u) = F(B\,u ,\ B\,u) .$$

Proof. Every mapping (17) is composed of a map (16) and of a map

$$\begin{cases} z \to A\ z \\ u \to B\ u \end{cases} \tag{19}$$

where A and B satisfy condition (18). It is clear that under this condition, (19) is an automorphism of the domain $D(V, F)$.

Suppose that we have an affine automorphism of the domain $D(V, F)$

$$\begin{cases} z \to L_{11}\ z + L_{12}\ u + b_1 , \\ u \to L_{21}\ z + L_{22}\ u + b_2 . \end{cases} \tag{20}$$

Combining (20) with a mapping (16), we can arrange that $b_2=0$, $\operatorname{Re}\ b_1 = 0$. We shall consider mappings of this form.

Furthermore the map (20) must preserve the skeleton Ω_D.

Therefore the point $(0, 0)$ must be transformed into a point of the skeleton, i.e. $\operatorname{Im}\ z = F(u, u)$. It follows that $b_1 = 0$, since $(0, 0) \to (b_1, 0)$, $\operatorname{Re}\ b_1 = 0$. So we may assume that in (20) $b_1 = b_2 = 0$.

Consider the points $(x, 0)$, $x \in \mathbb{R}^n$. They belong to Ω_D, and therefore their image belongs to Ω_D, i.e.

$$\operatorname{Im}\ L_{11} = F(L_{21}\ x,\ L_{21}\ x)\ .$$

Since the left-hand side is a linear form, and the right-hand side is a quadratic form, we have

$$\operatorname{Im}\ L_{11} = 0\ ,\quad L_{21} = 0\ .$$

Consider the points (iy, u), $y = F(u, u)$. This images must belong to the skeleton; i.e.

$$L_{11}\ y + \operatorname{Im}\ L_{12}\ u = F(L_{22}\ u,\ L_{22}\ u)\ . \tag{21}$$

Whence $\operatorname{Im} L_{12} e^{i} u$ is independent of , i.e. $L_{12} u = 0$, and , since this is true for every u, then

$$L_{12} = 0 .$$

By (21) , furthermore

$$L_{11} y = L_{11} F(u,u) = F(L_{22} u, L_{22} u) ,$$

i.e. our map has the form (19) , and A, B satisfy (18) . The proof is complete.

The V-hermitian form F is called <u>homogeneous,</u> if there exists a transitive group G of automorphisms of V such that for every $g \in G$ there exists a linear transformation $\tilde{g}$ of the space $\mathbb{C}^m$ such that

$$g\, F(u,v) = F(\tilde{g}\, u, \tilde{g}\, v) . \tag{22}$$

<u>Corollary</u> 1. The Siegel domain of type II D(V, F) is homogeneous if and only if V is a homogeneous cone and the V - hermitian form F is homogeneous .

For this it is enough to remark that, by a map (16) the point $(z,u) \in D(V,F)$ can be transformed into a point $(i\, y, 0), y \in V$ by proposition 4 . These points must be transformable into each other by maps of the form (19) .

Proposition 4 has the following generalization :

<u>Proposition</u> 5. Every non singular affine transformation mapping the Siegel domain of type II D(V, F) out a Siegel domain of type II $D(V_1, F_1) \subset \mathbb{C}^{n_1 + m_1}$ is of the form

$$\begin{cases} z \to Az + a + 2i\, F_1(Bu, c) + i\, F_1(c,c) , \\ u \to Bu + c, \end{cases}$$

where $a \in \mathbb{R}^{n_1}$, $c \in \mathbb{C}^{m_1}$, A is a linear transformation of the cone

V onto V_1, B is a linear transformation of $\mathbb{C}^m$ into $\mathbb{C}^{m_1}$ such that

$$A(F(u,u)) = F_1(Bu, Bu) .$$

The proof is analogous to the proof of proposition 4.

Corollary. The Siegel domains of type II $D(V,F) \subset \mathbb{C}^{n+m}$ and $D(V_1, F_1) \subset \mathbb{C}^{n_1+m_1}$ are affine equivalent if, and only if, $n = n_1$, $m = m_1$ and there exist isomorphic linear transformations $A : \mathbb{R}^n \to \mathbb{R}^{n_1}$, $B : \mathbb{C}^m \to \mathbb{C}^{m_1}$ such that

$$A\,V = V_1 ,$$
$$A\,F(u,u) = F_1(Bu, Bu) .$$

The study of homogeneous Siegel domains of type II is reduced to the study of homogeneous V - hermitian forms for homogeneous cones V . The classification of these forms up to linear equivalence for concrete cones is an interesting problem of linear algebra. Let us consider the Siegel domains associated with the cones of example 1 and 2.

Example 4. Let V be the cone of symmetric positive definite matrices of order ℓ. It can always be assumed (cf. the following paragraph) that the map (3), where g is upper triangular matrix with positive diagonal elements, can be continued to $\mathbb{C}^m$, in the sense of (22). Let us first consider the case where $\mathbb{C}^m$ is a space of rectangular $\ell \times q$ matrices u .

We set

$$F(u,v) = \frac{1}{2}(u\,v^* + \bar{v}\,u') . \tag{23}$$

It is clear that F is a V-hermitian form. Let us show that is homogeneous. We consider the maps of the cone V :

$$y \to g(y) = t\, g\, t$$

where t is an upper triangular matrix with positive diagonal elements. For $u \in \mathbb{C}^m$ we set

$$\tilde{g}(u) = t\,u. \tag{24}$$

For these maps conditions (22) is satisfied.

We obtain other examples of homogeneous V-hermitian forms if we restrict the form (23) to subspaces $\mathbb{C}^{m_o}$ of the space $\mathbb{C}^m$ which are invariant under left multiplication by upper-triangular matrices t (mappings (24)). For this it is necessary that the rows $u_1, \dots, u_\ell$ of the matrix u belong to subspaces $\mathbb{C}^{q_1}, \dots, \mathbb{C}^{q}$ of some chain of subspaces $\mathbb{C}^{q_1} > \dots > \mathbb{C}^{q}$ of the space $\mathbb{C}^q$. It is clear that one can choose a basis in $\mathbb{C}^q$ so that the subspace $\mathbb{C}^{m_o}$ consists of step matrices of some type. (The first elements in each row vanish, and the number of these elements does not decrease when we go from one row to the following).

It turns out that the domains D(V, F) associated to the form (23) are non-symmetric unless D(V, F) is a Siegel domain of type I. We prove this in the simplest case.

Let $\ell = 2$, $q = 1$, let u be matrices with the second row equal to zero. Then we obtain the domain in $\mathbb{C}^4$ given by the following conditions:

$$\begin{vmatrix} y_{11} - |u|^2 & y_{12} \\ y_{12} & y_{22} \end{vmatrix} > 0,\ y_{11} - |u|^2 > 0 \quad (y_{ij} = \mathrm{Im}\, z_{ij}). \tag{25}$$

This is the first example of a non-symmetric homogeneous bounded domain constructed by I.I.Pjateckii-Šapiro in [16]. We give a proof that it is not symmetric.

First of all we prove the following lemma:

<u>Lemma</u> 3. The symmetry of the Siegel domain of type II D(V, F)

at the point $(z_0, 0)$, if it exists, is of the form

$$(z, u) \longrightarrow (\varphi(z), \psi(z)u)$$

where $z \longrightarrow \varphi(z)$ is the symmetry of the Siegel domain of type I $D(V)$ at the point z_0, and $\psi(z)$ is a linear transformation of $\mathbb{C}^m$ depending analytically on z.

Proof. We mention first that the symmetry is unique at every point (it must be the reflection in the geodesies with respect to the Bergman metric). Let the symmetry at the point $(z_0, 0)$ be

$$(z, u) \rightarrow (\varphi(z, u), \psi(z, u)) . \tag{26}$$

It must commute with every automorphism of $D(V, F)$ which preserves the point $(z_0, 0)$ (because of uniqueness), in particular with

$$(z, u) \rightarrow (s, e^{i\theta} u) .$$

Hence

$$\varphi(z, e^{i\theta} u) = \varphi(z, u) ,$$

$$\psi(z, e^{i\theta} u) = e^{i\theta} \psi(z, u) .$$

Because of the analyticity of φ and ψ in a neighborhood of 0 with respect to u we obtain that the symmetry has the form (26). Setting $u = 0$, we obtain that $z \longrightarrow \varphi(z)$ is the symmetry of the domain $D(V)$ at the point z_0.

Lemma 4. The domain (25) is non-symmetric.

Proof. The symmetry at the point $(z = iE, u = 0)$ must be of the form

$$(z, u) \rightarrow (-z^{-1}, \psi(z)u) . \tag{27}$$

Under an analytic automorphism a point of the skeleton Ω_D must go into another point of the skeleton or to infinity.

However under the map (27) the point of the skeleton $z = \begin{pmatrix} i & 1 \\ 1 & 1 \end{pmatrix}$, $u = 1$ goes into the point $z = \frac{1}{2}\begin{pmatrix} 1+i & -1-i \\ -1-i & -1+i \end{pmatrix}$ which does not belong to the skeleton. Therefore there exists no symmetry at the point $(i\,E, 0)$.

Remark. It would be possible to compute the volume element for the Bergman metric of the domain (26), and check that it is not invariant under maps of the form (27).

Example 5. Let V be the cone of hermitian positive definite matrices of order ℓ. We realize the space $\mathbb{C}^m$ as the space of pairs of complex rectangular matrices $u^{(1)}$ of type $\ell \times q$ and $u^{(2)}$ of type $(\ell \times r)$. We set

$$(28) \qquad F(u, v) = u^{(1)}\, v^{(1)*} + \overline{v^{(2)}}\, u^{(2)} \quad .$$

Let t be an upper triangular (complex) matrix of order ℓ. To the automorphisms of the cone V

$$v \to g(y) = t\, y\, t^*$$

we make correspond the map of $\mathbb{C}^m$

$$(29) \qquad \tilde{g}(u^{(1)}, u^{(2)}) = (t\, u^{(1)}, \bar{t}\, u^{(2)}) \quad .$$

Condition (22) is satisfied. The corresponding domains are symmetric if one of the number q, r is equal to zero.

We denote by u_k the pair $(u_k^{(1)}, u_k^{(2)})$ consisting of the k-th rows of the matrices $u^{(1)}, u^{(2)}$; $u_k \in \mathbb{C}^{q+r}$. If we choose a chain of subspaces $\mathbb{C}^{s_1} \supset \ldots \supset \mathbb{C}^{s}$ in $\mathbb{C}^{q+r}$, and consider the space $\mathbb{C}^{m_0}$ of pairs $u = (u^{(1)}, u^{(2)})$ for which $u_k \in \mathbb{C}^{s_k}$ $(k = 1, \ldots, \ell)$, then we obtain a subspace which is invariant under the maps (29). The restriction of the form (28) to $\mathbb{C}^{m_0}$ gives a homogeneous V - hermitian form. Among the forms so obtained there are families of non-equivalent forms depending on certain parameters. As a result, by proposition 5 we obtain a continuum of affinely non-isomorphic homogeneous Siegel domains of type II. By the results of the next part, they are also analytically non - isomorphic.

The simplest continuous family of non isomorphic domains is obtained for $\ell = 2$, $q = r = 1$, $s_1 = 2$, $s_2 = 1$. In this way we get a family of domains in $\mathbb{C}^7$ (cf. Part II). Besides this family, in $\mathbb{C}^7$ there are only finitely many analitically non isomorphic h.b.d.'s'..

We state now one of the fundamental theorems of the theory of h.b.d.'s in $\mathbb{C}^m$.

Theorem. Every homogeneous bounded domain in $\mathbb{C}^n$ is analytically isomorphic with a homogeneous Siegel domain of type II.

This theorem in its final form was proved by E.B. Vinberg, S.G. Gindikin and I.I. Pjatecckii-Šapiro [25].

In these lectures we will give a proof of it under certain hypotheses concerning the group of automorphisms of the domain.

3. The structure of the group of affine automorphisms of a homogeneous Siegel domain of type II.

Let $D = D(V, F)$ be a homogeneous Siegel domain of type II in C^{n+m}, let $G_a(D)$ be the group of affine automorphisms of D. It is clear that $G_a(D)$ is a closed subgroup in the group of all non-singular affine transformations of the space $\mathbb{C}^{n+m}$. In proposition 4 we found the general form of the transformations in $G_a(D)$. First of all we select the subgroup N(D) of transformations (16) of the group $G_a(D)$. It is immediately verified that N(D) is a normal nilpotent subgroup (of step 2) of $G_a(D)$. The maps of the form (19), i.e. the automorphisms of the cone V which can be continued to $\mathbb{C}^m$ in the sense of (22) form a complementary sub-group P(D) of N(D) in $G_a(D)$:

(30) $$G_a(D) = P(D) \circ N(D) .$$

We select in $G_a(D)$ the one-parameter subgroup $\{b_t\} \subset P(D)$ which will play an important role in our later considerations (cf. Part II)

(31) $$b_t : (z,u) \to (e^t z, e^{\frac{1}{2}t} u) .$$

The maps b_t commute with the maps in P(D). We compute their commutations with the elements of N(D). We set

(32) $$\tau(a) : (z,u) \quad (z+a, u) , \quad a \in \mathbb{R}^n$$

(33) $$\varphi(c) : (z,u) \to (z+2iF(u,u)+iF(c,c), \; u+c) , \quad c \in \mathbb{C}^m$$

Then

(34) $$b_t \; \tau(a) b_t^{-1} = \tau(e^t a) , \; b_t \; \varphi(c) \, b_t^{-1} = \varphi(e^{\frac{1}{2}t} c) .$$

Now we study the isotropy sub-group of the point $p \in D$.

<u>Lemma</u> 5. The isotropy sub-group K at any point $p \in D$ in $G_a(D)$ is a maximal compact subgroup of $G_a(D)$.

<u>Proof</u>. The compactness of the group K, which can be considered to be a linear group, follows from the fact that it preserves a riemmanian metric (the Bergman metric; cf. introduction).

Let now K_1 be some maximal compact subgroup, and M be a bounded open set contained in D. We consider the set $K_1 M$, which is the result of the application of the transformations from K_1 to the points of M ; $K_1 M$ is a bounded set. Its center of gravity p belongs to the domain D because D is convex, and is a fixed point of K_1, since the center of gravity is an affine invariant. Since the isotropy subgroup of the point p is compact, it coincides with K_1. It remains only to note that the isotropy subgroup of the various points are nonjugate, and the subgroups conjugate to K_1 are maximal compact.

From lemma 5 it follows in particular that the number of connected components of the group $G_a(D)$ is finite. In fact $G_a^o(D)$ - the connected component of the identity in $G_a(D)$ - must act transitively on D. Therefore every connected component of $G_a(D)$ will have non-empty intersection with K, and hence the number of components of $G_a(D)$ cannot be larger than the number of the components of the compact group K, that is finite.

<u>Lemma</u> 6. The group $G_a^o(D)$ coincides with the connected component of the identity of its own normalizer in the group $G_a(\mathbb{C}^{n+m})$ of all non singular affine transformations of the space $\mathbb{C}^{n+m}$.

<u>Proof.</u> Let N be the normalizer in question. It is clear that N contains $G_a(D)$. Let $p \in D$ and suppose that $g \in N$ differs sufficiently little from the identity map, so that $g\ p \in D$. Then

$$g\ D = g\ G_a^o(D)p = G_a^o(D)\ g\ p = D$$

i.e. $g \in G_a(D)$. Consequently in some neighborhood of the identity of $G_a(\mathbb{C}^{n+m})$ the subgroups $G_a(D)$ and N coincide and therefore their identity components coincide.

The subgroup $G \subset G_a(C^{n+m})$ is called <u>triangular</u> if the linear parts of the transformations in G can be expressed by triangular matrices in some basis. The affine group in C^{n+m} can be considered in a natural way as a linear group in $\mathbb{C}^{n+m+1}$.

Under this inclusion the triangularity of an affine group is equivalent with the triangularity of the corresponding linear group.

It is known [23] that the maximal connected triangular subgroups of any linear Lie group are conjugate. By the remark just made this theorem is also true for affine groups.

An affine group is said to be algebraic if it can be described as a subgroup of the full affine group whose coefficients satisfy certain polynomial equations. Again the algebraicity of an affine groups is equivalent to the algebraicity of the linear group corresponfin to it in the sense indicated above. Let G be the connected component of the identity of a lin linear algebraic group. Then [23] it admits the factorization formula

(35) $$G = K \; T$$

where K is a connected compact subgroup and T is a connected triangular subgroup of the group G.

By the remarks made before the factorization (35) is valid also for affine groups. Let us study this factorization. First of all the subgroup K and T intersect only in the identity element (K consists of semi-simple elements with eigen-values of modulus one, while the elements of T have positive eigen-values). From this considerations it also follows that K is a maximal compact subgroup and T is a maximal connected triangular subgroup. Finally it can be immediately verified that K and T in (35) may be replaced by any subgroup conjugate to them, i.e. we can take any maximal compact subgroup and any connected triangular subgroup.

Proposition 6. The group $G_a(D)$ for a homogeneous Siegel domain D of type II coincides with the connected component of the identity of some affine algebraic group.

Proof. We prove that the normalizer N of the group $G_a(D)$ is an algebraic group. Then proposition 6 will follow from lemma 6. The elements of the group N are given by the condition

$$g \quad G_a^o(D) \; g^{-1} \subset G_a^o(D) \quad .$$

Taking the adjoint representation a Ad of the group $G_a(\mathbb{C}^{n+m})$ in the Lie algebra of the corresponding affine group, this condition can be transcribed as

(36) $$(\mathrm{Ad} \;\; g) \; G_a(D) \subset G_a(D) \quad ,$$

where $G_a(D)$ is the Lie algebra of the group $G_a(D)$. This is to verify that the representation Ad is rational and therefore condition (36) is equivalent to a system of polynomial relations. From all this we obtain

Proposition 7. Let $G_a(D)$ be the group of affine automorphisms of a homogeneous Siegel domain of type II $D \subset \mathbb{C}^{n+m}$. Let $K_a(D)$ be the isotropy group of some point $p \in D$.

Let $T_a(D)$ be a maximal connected triangular sub-group of the group of the group $G_a(D)$. Then

(37) $$G_a(D) = K_a(D) \;\; T_a(D) \quad ,$$

where $K_a(D) \cap T_a(D) = \{e\}$. The group $T_a(D)$ acts without fixed points on the domain D.

This last statement follows from the fact that $T_a(D)$ intersects the isotropy sub-groups only in the identity element.

Remark. The nilpotent normal sub-group $N(D)$ is contained in $T_a(D)$. From proposition 4 it follows (cf. 30) that

(38) $$T_a(D) = T(V) \cdot N(D)$$

where the sub-group $T(V)$ is obtained by the extension in the sense (22) of some triangular group of automorphisms of the cone V.

4. Affine-homogeneous tube domain in $\mathbb{C}^n$.

The construction of the Siegel domains of type I admits the following generalization : Let $U \subset \mathbb{R}^n$ be an affine homogeneous domain. We consider the domain in $\mathbb{C}^n$

$$(39) \qquad D(U) = \mathbb{R}^n + i \; U .$$

It is clear that the affine automorphisms of the domain U can be continued to C^n and together with the translations of R^n they generate a transitive group of affine automorphisms of D(U). So D(U) is an affine homogeneous domain in $\mathbb{C}^n$.

It is usual to call the domains of the form (39) tube domains with base U.

By a theorem of Bochner the tube domain D(U) is a domain of holomorphy if and only if its base U is convex.

Proposition 8. The tube domain D(U) is analytically isomorphic with a bounded domain if and only if the convex hull $\tilde{U}$ of its base does not contain any straight line.

The sufficiency of the condition is proved in the same way as in proposition 1. The necessity follows from the fact that if $\tilde{U}$ contains a straight line, the domain D(U) will contain a complex line. Then, by Lewy's theorem, every bounded holomorphic function will have to be constant on this line.

From proposition 8 it follows immediately that

Proposition 9. If $U \subset R^n$ is an affine homogeneous domain such that the domain D(U) is analytically isomorphic with a bounded domain, then U is a convex domain not containing any straight line.

In particular, if U is an affine homogeneous domain in R^n the convex hull of which contains no straight line, then U is a convex domain.

To prove the sufficiency, it is enough to note that an H.B.D. is a domain of holomorphy and to use the theorem of Bochner quoted in proposition 8 .

Now we formulate a result concerning the conditions that a homogeneous Siegel domain of type II be analytically isomorphic with a tube domain .

Proposition 10 . The homogeneous Siegel domain of type II D(V, F) is analytically isomorphic with a tube domain D(U) if and only if for an appropriate choice of a real sub-space $\mathbb{R}^m$ in V^m we have

$$F(u, v) \in \mathbb{R}^n \quad \text{for all} \quad u, v \in \mathbb{R}^m \quad . \tag{40}$$

In this case one can take for U the domain in $\mathbb{R}^{n+m}$

$$y - F(u, u) \in V \qquad (y \in \mathbb{R}^n, \ u \in \mathbb{R}^m) \ . \tag{41}$$

We shall not show here the necessity of these conditions, since this would require a long analysis of the group of analytic automorphisms of the domain D(V, F) .

The sufficiency is very easy to prove.

We make the analytic transformation

$$z \longrightarrow z - i \ F(u, \bar{u}) \ , \qquad u \rightarrow \sqrt{2} \ u$$

and as a result we obtain the tube domain with basis (41) .

We mention that it can be deduced from proposition 10 that every real convex affine homogeneous domain containing no striaght line is affinely equivalent with a domain of the form (41) . However this result can also be obtained directly [10, 24] .

The domain of example 4 satisfies the condition (40) , and therefore is analytically isomorphic with a tube domain. The domain of example 5 cannot be mapped onto a tube domain.

PART II - Linearization.

1. Kähler algebras.

The fundamental method in the theory of h.K.m.'s (just as in the theory of Lie groups and of homogeneous spaces) is the reduction to Lie algebras. We shall show here that to every h.K.m. M it is possible to make correspond a Lie algebras with certain additional structures, which determines M uniquely up to local isomorphisms. Let G be a connected transitive group of automorphisms of the manifold M . Let K be the isotropy sub-group at the point $p \in M$. We denote by $\mathfrak{g}$ the Lie algebra of the group G and by $\mathfrak{K}$ its sub-algebra corresponding to K . We consider the map

$$\pi : \quad G \longrightarrow M$$

defined by the formula

$$\pi(g) = g(p) \; . \tag{1}$$

Its differential $d\pi$ at the point $e \in G$ is a linear transformation of the Lie algebra $\mathfrak{g}$ onto the tangent space T of the manifold M at the point p . The kernel of $d\pi$ is $\mathfrak{K}$. We define a linear operator j on $\mathfrak{g}$ such that

$$d\pi(jx) = I \; d\pi(x) \quad , \tag{2}$$

where I is the operator of the complex structure in T . The operator j is determined in this way only modulo π .

We define furthermore the antisymmetric bilinear form ϱ on $\mathfrak{g}$ by the formula

$$\varrho(x, y) = \eta(d\pi(x) \, , \; d\pi(y)) \tag{3}$$

where η = Im h (see § 1 of the introduction) . The quadruple $\{\mathfrak{g}, \pi, j, \varrho\}$ defined in this way has the following properties :

(K A 1) $\quad j\mathcal{K} \subset \mathcal{K}, \quad j^2 \equiv -1 \pmod{\mathcal{K}}$;

(K A 2) $\quad [k\ jx] \equiv j\,[k, x] \pmod{\mathcal{K}}$ for $k \in \mathcal{K}$;

(K A 3) $\quad [jx, jy] \equiv j\,[jx, y] + j\,[x, jy] + [x, y] \pmod{\mathcal{K}}$;

(K A 4) $\quad \varrho(k, x) = 0$ for $k \in \mathcal{K}$;

(K A 5) $\quad \varrho(jx, jy) = \varrho(x, y)$;

(K A 6) $\quad \varrho(jx, x) > 0$ for $x \notin \mathcal{K}$;

(K A 7) $\quad \varrho([x, y], z) + \varrho([y, z], x) + \varrho([z, x], y) = 0$.

The property (K A 1) follows immediately from the definition of j and from the fact that $j^2 = -1$. The property (K A 2) expresses the invariants of the complex structure on T with respect to the isotropy group. To prove it one has to use the following well known fact:

If $d\,k$ is the differential of the map $k \in K$ at the point p, then

(4) $$dk(d\pi(x)) = d\pi((\mathrm{Ad}\,k)\,x) .$$

The property (K A 3) is the integrability condition of the complex structure on M (see e.g. [9]).

The property (K A 4) follows immediately from the definition of ϱ.

The properties (K A 5) and (K A 6) express the fact that η is the imaginary part of a positive definite hermitian form.

Finally property (K A 7) means, in virtue of a known formula, that the left invariant differential form on G which is equal to ϱ at the identity, is closed, which fact in turn is equivalent to the closure of the form η. In this way (K A 7) is the Kähler condition (K 1) (see § 1 of the introduction) transcribed in terms of the Lie algebra $\mathfrak{g}$.

The quadruple $\{\mathfrak{g}, \pi, j, \varrho\}$ consisting of a Lie algebra $\mathfrak{g}$, a sub-algebra $\mathcal{K}$, a linear operator j, and a antisymmetric bilinear form ρ on $\mathfrak{g}$, will be called a Kähler algebra if it has properties (K A 1) - (K A 7). By abuse of language we will often call the Lie algebra $\mathfrak{g}$ a Kähler algebra.

By construction the group G acts effectively on M. This however need not always be so. If it is true, then the Kähler algebra $\{\mathfrak{g}, \mathcal{K}, j, \rho\}$ also has the following property ; $(\rho) \in \kappa$ contains no non-zero ideal of G.

A Kähler algebra satisfying this condition will be called <u>effective.</u>

From the invariance of the function with respect to the isotropy group, it follows that

(5) $$\rho([k,x], y) + \rho(x, [k,y]) = 0 \quad \forall k \in \mathcal{K}.$$

However this condition is not new. It is a consequence of (K A 7) if we set $z = k$ and use (K A 4).

In this way to every h.K.m. M together with a transitive group of automorphisms G acting on it, we have associated an effective Kähler algebra.

It is easy to see that this Kähler algebra determines the Kähler manifold M and the group G uniquely up to local isomorphisms.

Conversely, let $\{\mathfrak{g}, \mathcal{K}, j, \rho\}$ be an arbitrary Kähler algebra and let G be any connected Lie group having $\mathfrak{g}$ as its Lie algebra. We assume that the connected sub-group $K \subset G$ corresponding to the sub-algebra $\mathcal{K} \subset \mathfrak{g}$, is closed. Then on the homogeneous space $M = G/K$ there is a uniquely determined invariant Kähler structure associated to j and ρ by the relations (2) and (3). The Kähler algebra constructed from $\{M, G\}$ will coincide with the original $\{\mathfrak{g}, \mathcal{K}, j, \rho\}$.

2. Kähler algebras corresponding to the three fundamental types of homogeneous Kähler manifolds.

For each of the three fundamental types of (h.K.m.) (see Introduction) it is possible to establish conditions which characterize the Kähler algebras corresponding to the h.K.m. of that type .

We give here these conditions and prove their necessity. The sufficiency, except for the trivial case of a locally flat h.K.m. , can be obtained from the classification of (h.K.m.) of the corresponding types [14 , 25] .

Let M be a locally flat h.K.m., and let G be an arbitrary group of its automorphisms containing the group I of parallel translations. The group G can be factored into a semidirect product

$$G = K.I$$

and its Lie algebra $\mathcal{G}$ into a semidirect sum

(6) $$\mathcal{G} = \mathcal{K} + \mathcal{J}$$

where $\mathcal{J}$ is a commutative ideal corresponding to I .

Conversely if the Kähler algebra $\mathcal{G}$ corresponding to the h.K.m. M admits a factoring (6) , then M is a locally flat h.K.m. since the commutative group I corresponding to the ideal $\mathcal{J}$ acts on it transitively.

Let us consider now the other two fundamental types of h.K.m. The Kähler algebras corresponding to these types of h.K.m. have one common property

(ND) $$\varphi(x, y) = \omega([x, y])$$

where ω is some linear function on $\mathcal{G}$.

The Kähler algebras having this property will be called <u>non-degenerate</u> (in [17 - 21, 25] they were called j - algebras) .

The non-degeneracy means that the form φ is cohomologous to zero in the sense of the cohomology of the Lie algebra $\mathcal{G}$. In this case it is automatically closed, i.e. the condition (K A 7) is satisfied.

If $\mathcal{G}$ is a semi-simple Lie algebra then $H^2(G) = 0$. Therefore all semi--simple Kähler algebras are non-degenerate.

In Koszul's work [9] it is shown that if the coefficients of the hermitian form h, defining the Kähler structure on the h.K.m., are of the form

$$h_{\alpha\beta} = \frac{\partial \log \varphi}{\partial z^{\alpha} \partial \bar{z}^{\beta}} ,$$

where φ is the density of the invariant measure, then the corresponding Kähler algebra is non-degenerate.

In particular this is true for h.b.d.'s having a Bergman metric (using the classification of h.b.d.'s [25] one can show the non-degeneracy of the Kähler algebra corresponding to a h.b.d. with any Kähler structure, cf. lemma 1 in §1 of part 3).

Thus the Kähler algebras corresponding to simply connected compact h.K.m.'s and also those corresponding to h.b.d.'s are non-degenerate. On the other hand from the decomposition (6) one sees that in a Kähler algebra corresponding to a locally flat h.K.m., every skew symmetric bilinear function ρ having properties (ND) and (K A 4), is zero.

In this sense these Kähler algebras are "maximally degenerate".

In order to get a property which distinguishes the Kähler algebras corresponding to simply connected compact h.k.m.'s from the Kähler algebras corresponding to h.b.d.'s we introduce the notion of a Kähler sub-algebra.

Let $\{\mathcal{G}, \mathcal{K}, j, \rho\}$ be a Kähler algebra and let $\mathcal{G}_1$ be a sub-algebra of $\mathcal{G}$ satisfying the condition

(7) $$j\, G_1 \subset G_1 + K .$$

On the space $\mathcal{G}_1$ one can define an operator j_1 such that $j\,x \equiv j_1\, x \pmod{\mathcal{K}}$ for all $x \in \mathcal{G}_1$.

We define the bilinear form $\mathfrak{g}_1$ on $\mathfrak{g}_1$ as the restriction of the form $\mathfrak{g}$ and we set $\mathcal{K}_1 = \mathcal{K} \cap \mathfrak{g}_1$. It is easy to see that $\{\mathfrak{g}_1, \mathcal{K}_1, j_1, \rho_1\}$ is a Kähler algebra. It is called a Kähler sub-algebra $\{\mathfrak{g}, \mathcal{K}, j\ \rho.\}$

The Kähler algebra $\{\mathfrak{g}, \mathcal{K}, j, \rho\}$ is called proper if it has the following property :

(R) Every compact semi-simple Kähler sub-algebra of $\mathfrak{g}$ is contained in K .

It is clear that the Kähler algebra corresponding to a simply connected compact h.K.m. is "maximally non-proper" since it is itself semi-simple and compact.

On the other hand the Kähler algebras corresponding to h.b.d.'s are proper.

In fact let $\{\mathfrak{g}, \mathcal{K}, j, \rho\}$ be the Kähler algebra corresponding to the h.b.d. D with the transitive group of automorphisms G, and let $\mathfrak{g}_1$ be a compact semi-simple Kahler sub-algebra in it.

We denote by G_1 the connected compact sub-group of G corresponding to $\mathfrak{g}_1$ and we consider the orbit $G_1 p$ of the point $p \in D$ (the isotropy group of which is K) . This orbit is a compact complex sub-manifold of D. The coordinate functions restricted to $G_1 p$ must be constant, therefore $G_1 p = \{p\}$. This means that $G_1 \subset K$ and $\mathfrak{g}_1 \subset \mathcal{K}$ which has to be proved.

§ 3. Kähler algebras corresponding to homogeneous Siegel domains.

Let us consider the structure of a Kähler algebra corresponding to a homogeneous Siegel domain of type II D(V, F) associated to the convex cone $V \subset \mathbb{R}^n$ and to the V-hermitian form F in $\mathbb{C}^m$.

First of all we describe the infinitesimal affine transformations corresponding to the affine automorphisms of the domain $D(V,F)$.

This can be obtained from the formula (17) of the first part by differentiating with respect to the parameters of the group. We obtain the following infinitesimal affine transformations

(8) $$(x+iy, u) \to (Ax+iAy,\ Bu)\ ,$$

(9) $$(x+iy, u) \to (a, 0) \qquad (a \in \mathbb{R}^n)\ ,$$

(10) $$(x+iy, u) \qquad (2i\, F(u,c), c) \quad (c \in \mathbb{C}^m)\ .$$

Where in (8) A is an element of the Lie algebra of the automorphism group of the cone V and B is a linear transformation of $\mathbb{C}^m$ related to A by the condition

$$AF(u,v) = F(Bu,v) + F(u,Bv)\ .$$

To the one-parameter group $\{b_t\}$ (see the first part) there corresponds the infinitesimal transformation

(12) $$g_o : (x+iy, u) \to (x+iy, \tfrac{1}{2}\, u)\ ,$$

so that

$$b_t = \exp\ t\ g_o\ .$$

Let now G be any transitive group of affine automorphisms of the domain $D(V,F)$ containing all maps (16) of the first part and also the one parameter group $\{b_t\}$. We denote by $\mathfrak{g}$ its Lie algebra and by $\mathfrak{g}_o$, $\mathfrak{g}_1$ and $\mathfrak{g}_{\frac{1}{2}}$ the sub-spaces of the algebra $\mathfrak{g}$ corresponding to the transformations (8), (9) and (10) respectively.

It is clear that

$$\mathfrak{g} = \mathfrak{g}_o + \mathfrak{g}_1 + \mathfrak{g}_{\frac{1}{2}}$$

and $g_0 \in \mathfrak{g}_0$. Taking the commutator of g_0 with the infinitesimal transformations (8), (9) and (10) we see that $\mathfrak{g}_\lambda (= 0, 1, 1/2)$ is the eigenspace of ad $\mathfrak{g}_0$ corresponding to the eigenvalue λ. Whence in particular it follows that

(14) $$[\mathfrak{g}_\lambda, \mathfrak{g}_\mu] \subset \mathfrak{g}_{\lambda+\mu} .$$

In order to define to operator j in the algebra $\mathfrak{g}$ we have to pick a base point p D(V, F). We choose

$$p = (i\mathfrak{y}_0, 0)$$

where $\mathfrak{y}_0$ is some point of the cone V.

Computing j by the formula (2) we find that

(15) $$j\mathfrak{g}_\lambda \subset \mathfrak{g}_{1-\lambda}$$

(more exactly one can arrange this to be true using the arbitrariness in the definition of j).

Since all affine automorphisms of D(V, F) preserving the point p are of the form (19) of part I, we have $\kappa \subset \mathfrak{g}_0$.

Lemma. The element g_0 of the Kähler algebra $\mathfrak{g}$ is uniquely determined by the following conditions :

1) The operator ad g_0 is semi-simple and its eigenvalues are 0, 1, $1/2$

2) Let $\mathfrak{g}_\lambda$ be the eigenspace of ad g_0 corresponding to the eigenvalue λ; then for appropriate choice of j

$$j\mathfrak{g}_\lambda \subset \mathfrak{g}_{1-\lambda} ;$$

3) $$\mathcal{K} \subset \mathfrak{g}_0 .$$

Proof. Let $\tilde{g}_0$ be another element possessing the properties listed in the lemma. From these properties it follows that if the subspace $\mathcal{W} \subset \mathcal{G}$ is invariant under ad $\tilde{g}_0$ then also the subspace $j\mathcal{W} \in \mathcal{K}$ will be invariant (for any choice of j) .

We keep the notation $\mathcal{G}_\lambda$ for the eigenspaces of ad g_0. By (14) the subspace $\mathcal{G}_1$ is an ideal and consequently it is invariant under ad g_0.

Therefore also $\mathcal{G}_0$ is invariant under ad $\tilde{g}_0$. In particular $[g_0, \tilde{g}_0] \in \mathcal{G}_0$. Hence it follows that $\tilde{g}_0 \in \mathcal{G}_0$, but then $j\tilde{g}_0 \in \mathcal{G}_1$, i.e. $[g_0, j\tilde{g}_0] = j\tilde{g}_0$.

Since g_0 and $\tilde{g}_0$ can be interchanged except for the choice of j, which depends on g_0, we have $[\tilde{g}_0, j\tilde{g}_0] \equiv j g_0 \pmod{\mathcal{K}}$.

Now we write down the integrability condition (K A 3) for the elements g_0 and $\tilde{g}_0$:

$$[j g_0, j\tilde{g}_0] \equiv j[j g_0, \tilde{g}_0] + j[g_0, j\tilde{g}_0] + [g_0, \tilde{g}_1] \pmod{\mathcal{K}}$$

whence

$$\tilde{g}_0 \equiv g_0 \pmod{\mathcal{K}} .$$

The precise equality of $\tilde{g}_0$ and g_0 follows from the reality of their eigenvalues. The lemma is proved.

We have

$$(16) \qquad g_0 = j\,s \quad \text{where} \quad s \in \mathcal{G}_1 .$$

The element is called the <u>principal idempotent</u> of the Kähler algebra $\mathcal{G}$.

The corresponding infinitesimal transformation is of the form

$$(17) \qquad s : (x + i\,y,\ u) \rightarrow (y_0\ ,0) .$$

<u>Proposition 1</u> : The homogeneous Siegel domain of type II D(V, F) is uniquely determined by its Kähler algebra which is constructed from a transitive group of affine automorphisms containing all transformations (16) of part 1 and the one-parameter group $\{b_t\}$.

Proof : We show how the cone V and the form F can be reconstructed from the algebra $\mathfrak{g}$. To each $a \in \mathbb{R}^n$ we make correspond the infinitesimal transformation (9) .

In this way we get an isomorphic map of $\mathbb{R}^n$ onto the space $\mathfrak{g}_1$. We denote it by τ. Formula (17) shows that $\tau(y_0) = S$. We denote by S_{AB} the infinitesimal transformation (8) . An immediate computation shows that

$$[S_{AB}, \tau(a)] = \tau(Aa) \tag{18}$$

therefore the diagram

$$\begin{array}{ccc} \mathbb{R}^n & \xrightarrow{\quad A \quad} & \mathbb{R}^n \\ {\scriptstyle\tau}\downarrow & \text{ad } S_{A,B} & \downarrow{\scriptstyle\tau} \\ \mathfrak{g}_1 & \xrightarrow{\qquad\quad} & \mathfrak{g}_1 \end{array} \tag{19}$$

is commutative.

The cone V is the orbit of y_0 for the action of the group generated by the infinitesimal transformations A .

Consequently the cone $\tau(V)$ is the orbit of the point $\tau(y_0) = S$ under the group generated by the infinitesimal transformations ad S_{AB}.

This determines a cone $\tau(V)$ and therefore also the cone V isomorphic with it in terms of the algebra $\mathfrak{g}$. Making correspond to every $c \in \mathbb{C}^m$ the infinitesimal transformation (10) we obtain an isomorphic map of the vector space C^m onto the space $\mathfrak{g}_{\frac{1}{2}}$. We denote it by φ .

Computing some commutators we immediately see that

$$[\varphi(c_1), \varphi(c_2)] = \tau(2iF(c_2, c_1) - 2iF(c_1, c_2)) = 4\tau(\text{im } F(c_1, c_2)) . \tag{20}$$

From the hermiticity of the form F it follows that

(21) $$F(c,c) = \operatorname{Im} \; F(ic,c)$$

so that

(22) $$\tau(F(c,c)) = \frac{1}{4}[j\varphi(c), \varphi(c)] \; .$$

We have a commutative diagram

(23) $$\begin{array}{ccc} C^m & \xrightarrow{F} & \mathbb{R}^n \\ \downarrow \tau & & \downarrow \tau \\ \mathfrak{g}_{\frac{1}{2}} & \longrightarrow & \mathfrak{g}_1 \end{array}$$

where the map at the bottom is defined by the formula

(24) $$u \rightarrow \frac{1}{4}[j\,u\,,\,u] \; .$$

This determines F in terms of the algebra $\mathfrak{g}$.

Proposition 2 : Let $\mathfrak{g}$ be a non-degenerate Kähler algebra having an element s with the following properties :

1) $[\mathcal{X}, s] = [\mathcal{X}, js] = 0$;

2) $[js, s] = s$;

3) The operator ad js is semi-simple and has eigenvalues $0, 1$ and $1/2$;

4) $j\,\mathfrak{g}_\lambda \subset \mathfrak{g}_{1-\lambda}$ where $\mathfrak{g}_\lambda$ is the eigenspace of ad js corresponding to the eigenvalue λ .

Then $\mathfrak{g}$ is the Kähler algebra corresponding to some Siegel domain of type II $D = D(V,F)$ with a transitive group of affine automorphisms satisfying the conditions of propositions 1 .

The domain D is a Siegel domain of type I if and only if the operator ad js does not have $\frac{1}{2}$ eigenvalue.

We mention that condition 2) follows from condition 4) since $js \in \mathfrak{g}_0$.

The condition of non-degeneracy of the algebra $\mathfrak{g}$ can also be omitted.

For this see lemma 1 in §1 of part 3 .

The way of constructing the Siegel domain D(V, F) which has $\mathfrak{g}$ as its Kähler algebra is clear from the proof of proposition 1. As V one has to take the orbit of the point $s \in \mathfrak{g}_1$ under the group generated by the infinitesimal transformations ad X , $x \in \mathfrak{g}_0$. The form F is defined by formula (24) .

It is necessary to prove also that V is a convex cone and F is a V - hermitian form . This is the hardest part of the proof of proposition 2 . We will not carry it out here . It can be found in paragraph 1 of our article [25] .

We mention, however, that if it is known a priori that $\mathfrak{g}$ is the Kähler algebra of some h.b.d. then the convexity of the cone V and the V - hermiticity of the form F can be obtained from analytic considerations (cf. part 1) .

The verification that $\mathfrak{g}$ is in fact the Kähler algebra of the Siegel domain just constructed is not difficult.

The last statement of proposition 2 is obvious.

§4. Normal Kähler algebras.

In certain cases an h.K.m. admits a transitive splittable solvable group of automorphisms.

For example all homogeneous Siegel domains of type II have this property (cf. part 1) .

One can make use of this in order to remove the ambiguity in the construction of the Kähler algebra corresponding to a given h.K.m., which is there because of the several possibilities to choose the transitive group of automorphisms.

Proposition 3 : Let T be a transitive splittable solvable group of automorpphisms of the h.K.m. M. Then T acts on M without fixed points. All transitive splittable solvable groups of automorphisms of M are conjugate to T by an element of the isotropy group K of some fixed point p $\in$ M .

Proof. By hypothesis

$$G^{o}(M) = K T .$$

We show that

$$K \cap T = \{e\} .$$

We denote by $\mathfrak{g}$ (M) the Lie algebra of the group $G^{o}(M)$, and by $\mathcal{K}$ and $\mathcal{T}$ the Lie algebras of the groups K and T respectively.

We have :

$$\mathfrak{g}(M) = \mathcal{K} + \mathcal{T} .$$

Let $k \in K \cap T$. The operator induced by ad k in the space $\mathfrak{g}(M)/\mathcal{K} = \mathcal{T}/\mathcal{T} \cap \mathcal{K}$ preserves the positive definite hermitian form (cf. (4)). Consequently it is semi-simple and its eigenvalues have modulus 1 .

On the other hand from the splittability of the group T it follows that its eigenvalues are positive. From here it follows that ad k acts trivially on $\mathfrak{g}(M)/\mathcal{K}$, i.e. the map k of M has trivial linear part at p (cf. (4)) .

Since k is an isometry, k = e .

Thus $K \cap T = \{e\}$. If $\widetilde{T}$ is a splittable solvable sub-group of the group G(M) containing then, by the same argument, $K \cap \widetilde{T} = \{e\}$.

This is possible only if $\widetilde{T} = T$. Consequently T is a maximal splittable solvable sub-group of the group G(M) .

For the proof of the second assertion of proposition 3 it suffices to show that every transitive splittable solvable group $\widetilde{T}$ of automorphisms of M is conjugate to T by some element $g \in G^{o}(M)$.

In fact the element g can be represented in the form

$$g = k\,t \qquad (k \in K,\ t = T),$$

and then

$$\widetilde{T} = g\,T\,g^{-1} = k\,t\,T\,t^{-1}k^{-1} = k\,T\,k^{-1} .$$

We denote by Ad adjoint representation of the group $G^{o}(M)$. Its kernel is the center Z of the group $G^{o}(M)$.

It is obvious that every maximal splittable solvable sub-group of the group $G^{o}(M)$ must contain the connected component Z^{o} of Z.

Therefore in order to show that T and $\widetilde{T}$ are conjugate in $G^{o}(M)$ it suffices to show that ad T and ad $\widetilde{T}$ are conjugate in ad $G^{o}(M)$.

We make use now of the following theorem (cf.e.g. [23]).

In every linear Lie group all maximal connected triangular subgroups are conjugate.

In order to apply this theorem we have to show that the group Ad T is triangular, i.e. that the operators Ad t, $t \in T$ have real eigenvalues not only on the algebra $\mathfrak{T}$ but also on the whole algebra $\mathfrak{G}(M)$.

From the fact that T is a maximal splittable solvable sub-group in G(M) (see above) it will then follows that Ad T is a maximal triangular sub-group in Ad $G^{o}(M)$. So it remains to prove that Ad T acting on the space $\mathfrak{G}(M)$ is triangular.

We do this only for the case when M is a h.b.d. In the general case the proof can be reduced to the case of a h.b.d. if one uses theorem 2, § 7.

If M is a h.b.d. then the group Ad $G^{o}(M)$ is the connnected component of the identity of some algebraic linear group. This is proved in § 3 of our article [25] .

From this it is easy to deduce the algebraicity of the group ad T . In fact ad T is a maximal splittable solvable sub-group in Ad $G^{o}(M)$. But its algebraic hull $(Ad\ T)_a$ (i.e. the smallest algebraic group containing Ad T) is also a splittable solvable group . Consequently $(Ad\ T)_a =$ $=$ Ad T .

Like every connected solvable algebraic linear group the group Ad T can be factored into a semi-direct product

$$Ad\ T = (Ad\ T)_R \cdot (Ad\ T)_I$$

where $(Ad\ T)_R$ is a normal sub-group containing all elements of Ad T which have positive eigenvalues and $(Ad\ T)_I$ is a commutative subgroup containing all semi-simple elements whose eigenvalues have modulus 1 . Since all linear transformations contained in Ad T have positive eigenvalues on $\mathcal{T}$ (by the fact that T is splittable) the linear transformations contained in $(Ad\ T)_I$ are equal to the identity on $\mathcal{T}$.

In other words all elements $t \in T$ for which $Ad\ t \in (Ad\ T)_I$ belong to the center of the group T .

Let us find the center of T . Since T is splittable it follows that it is connected. On the other hand the center of the algebra is trivial, since for any $x \in \mathcal{T}$, $x \neq 0$,

$$\omega([jx, x]) = \varrho(jx, x) > 0$$

and hence $[jx, x] \neq 0$. Therefore $(Ad\ T)_I = \{e\}$, and $Ad\ T = (Ad\ T)_R$, which we had to prove .

<u>Corollary 1</u> : If the h.K.m. M admits a transitive splittable solvable group of automorphisms , then all such groups determine isomorphic Kähler algebras.

In fact all these Kähler algebras are Kähler sub-algebras of $\mathcal{G}(M)$. By proposition 3 they are conjugate with respect to the group $Ad_{(M)} K$, which contains j and ϱ, and consequently they are isomorphic as Kähler algebras.

Corollary 2 : Let $D = D(V, F)$ and $D_1 = D(V_1, F_1)$ be homogeneous Siegel domains of type II. Then the following conditions are equivalent :

1) The domains D and D_1 are affine isomorphic.

2) The domains D and D_1 are analytically isomorphic.

3) The Kähler algebras constructed using transitive triangular groups of automorphisms for D and D_1 are isomorphic.

The implication 1)$\Rightarrow$2) is obvious. 2)$\Rightarrow$3) follows from corollary 1, and 3)$\Rightarrow$1) follows from proposition 1.

Now we introduce the following definition: A Kähler manifold will be called normal if it admits a transitive splittable solvable group of automorphisms.

The Kähler algebra $\{\mathcal{G}, \mathcal{K}, j, \varrho\}$ will be called normal if $\mathcal{K} = 0$ and $\mathcal{G}$ is a splittable solvable Lie algebra. If M is a h.K.m. and G is a transitive group of its automorphisms, then the Kähler algebra constructed from the pair $\{M, G\}$ will be normal if and only if G is a splittable solvable group.

By Corollary 1 of proposition 3, to every normal Kähler manifold there corresponds uniquley a normal Kähler algebra.

For normal Kähler algebras the axioms (K A 2) and (K A 4) are void. The remaining axioms have the following form :

(K A N 1) $\quad j^2 = -1$

(K A N 2) $\quad [jx, jy] = j[jx, y] + j[x, jy] + [x, y]$;

(K A N 5) $\quad \varrho(jx, jy) = \varrho(x, y)$;

(K A N 6) $\qquad \varrho(jx, x) > 0 \quad$ for $\quad x \neq 0$;

(K A N 7) $\qquad \varrho([x,y],z) + \varrho([y,z],x) + \varrho([z,x],y) = 0$.

5. Examples of normal Kähler algebras.

The simplest example of a normal Kähler algebra is a commutative Kähler algebra. It corresponds to a locally flat h.K.m.

A simple example of a non-degenerate normal Kähler algebra is the algebra

$$\mathcal{G} = (js) + (s) + \mathcal{U} \tag{24}$$

where s is the principal idempotent (cf. § 3), $\mathcal{U} = \mathcal{G}_{\frac{1}{2}}$ is a hermitian vector space. The commutator of the elements $u, v \in \mathcal{U}$ is given by the formula

$$[u, v] = (\mathrm{Im}\,(u, v))\, s \ . \tag{25}$$

The form ϱ is defined as $d\omega$, where ω is a linear form $= 0$ on $(js) + \mathcal{U}$ and positive on s .

By construction the Kähler algebra (24) has a principal idempotent and therefore corresponds to some homogeneous Siegel domain of type II D(V, F) .

Recalling the proof of proposition 2, it is easy to find V and F. As V we can take the positive half-line in the one dimensional space $(s) = \mathcal{G}_1$; then the form F will be given on $\mathcal{U}$ in the following way

$$F(u, u) = \frac{1}{4}\,[ju, u] = \frac{1}{4}\ \mathrm{Im}\,((ju, u))\, s = \frac{1}{4}\,(u, u) s$$

(see formulas (23) and (25)). In this way the domain $D(V, F)$ consists of the pairs (z, u) (z a complex number, $u \in \mathcal{U}$) satisfying the condition

$$\operatorname{Im} z - \frac{1}{4} (u, u) > 0 .$$

As we mentioned in the first part, the Siegel domain is analytically isomorphic with a complex ball.

The Kähler algebras of type (24) are called <u>elementary normal Kähler algebras.</u> It can be shown [19, 20] that every non-degenerate normal Kähler algebra can be decomposed into a semi-direct sum of elementary ones, i.e.

$$\mathcal{G} = \sum_{i=1}^{m} \mathcal{G}^{(i)} \tag{26}$$

where $\mathcal{G}^{(i)}$ is a Kähler sub-algebra which is an elementary normal Kähler algebra , and

$$\left[\mathcal{G}^{(i)}, \mathcal{G}^{(j)}\right] \subset \mathcal{G}^{(j)} \quad \text{for } i > j .$$

This decomposition is unique . The number m is called the rank of the algebra $\mathcal{G}$.

We describe the structure of non-degenerate normal Kähler algebras of rank 2 . Every such algebra decomposes into a direct sum of sub-spaces

$$\mathcal{G} = (js_1) + (s_1) + jX + X + U_1 + (js_2) + (s_2) + U_2 \tag{27}$$

where X is a Euclidean vector space, U_1, U_2 are hermitian vector spaces.

The operators ad js_1 and ad js_2 are scalar on each sub-space in the decomposition (27) . Their eigenvalues on these sub-spaces are given by the following table

(28)

0, 0	1, 0	$\frac{1}{2}, -\frac{1}{2}$	$\frac{1}{2}, \frac{1}{2}$	$\frac{1}{2}, 0$
		0, 0	0, 1	$0, \frac{1}{2}$

One sees from it that the element $s = s_1 + s_2$ is a principal idempotent and

$$\mathcal{G}_0 = (js_1) + (js_2) + jX \ ,$$
$$\mathcal{G}_1 = (s_1) + (s_2) + X \ ,$$
$$\mathcal{G}_{\frac{1}{2}} = \mathcal{U}_1 + \mathcal{U}_2 \ .$$

Furthermore we have the relations

$$[jX, jX] = [X, X] = 0 \ ,$$

$$[jx, y] = (x, y)s_1 \quad (x, y \in X) \ ,$$

(29) $$[u_k, v_k] = (\operatorname{Im}(u_k, v_k))s_k \quad (u_k, v_k \in \mathcal{U}_k, \ k = 1, 2) \ ,$$

$$[jx, s_2] = x \qquad (x \in X) \ .$$

The other commutators, except for $[jx, \mathcal{U}_2]$ and $[\mathcal{U}_1, \mathcal{U}_2]$ about which we shall speak below, are zero (this follows from the consideration of weights, cf. table (28)) .

The decomposition (26) has the form

$$\mathcal{G}^{(1)} = (js_1) + (s_1) + jX + X + \mathcal{U}_1$$
$$\mathcal{G}^{(2)} = (js_2) + (s_2) + \mathcal{U}_2 \ .$$

The form ϱ is defined as $d\omega$, where ω is a linear form equal to zero on each term of the decomposition (27) except (s_1) and (s_2)

and is positive on s_1 and s_2 .

Finally we describe the commutators on $jX \times \mathcal{U}_2$ and $\mathcal{U}_1 \times \mathcal{U}_2$. If until this moment the structure of the Kähler algebra $\mathcal{G}$ was determined only by the dimension of the spaces X, $\mathcal{U}_1, \mathcal{U}_2$ and by the positive numbers $\omega(s_1)$, $\omega(s_2)$, now in the definition of the commutators on $jX \times \mathcal{U}_2$ there is a great arbitrariness. Namely for $x \in X$ and $u \in \mathcal{U}_2$

(30) $$[jx, u] = x \circ u \in \mathcal{U}_1 ,$$

where the circle denotes an arbitrary bilinear map $X \times \mathcal{U}_2 \to \mathcal{U}_1$ having the following property

(31) $$(x \circ u, x \circ u) = \frac{1}{2} (x, x)(u, u) .$$

Commutation on $\mathcal{U}_1 \times \mathcal{U}_2$ is uniquely determined by the formula

(32) $$(x, [u_1, u_2]) = \operatorname{Im} (u_1, x \circ u_2), (x \in X) .$$

From (31) it follows that if $X \neq 0$ then $\dim \mathcal{U}_1 \geqslant \dim \mathcal{U}_2$ since for any x X, $x \neq 0$ the linear map $u \to x \circ u$ $(u \in \mathcal{U}_2)$ has a trivial kernel. The proof of these statements can be found in [20] .

§ 6 . Extension of normal Kähler algebras.

In a normal Kähler algebra there is a canonical positive definite scalar product

(33) $$(x, y) = \varrho (jx, y) .$$

Lemma 2 . Let $\mathcal{J}$ be a Kähler ideal of the normal Kähler algebra $\mathcal{G}$. Then the orthogonal complement $\mathcal{H}$ to $\mathcal{J}$ with respect to the canonical scalar product (33) is a Kähler sub-algebra.

Proof. Since $\mathcal{J}$ is an invariant sub-space under j, $\mathcal{H}$ coincides with the orthogonal complement to $\mathcal{J}$ with respect to the form ϱ and is also invariant under j . In order to prove that $\mathcal{H}$ is a sub-algebra it suffices now to write down the condition (K A 7) for $x, y \in \mathcal{H}$, z .

An analogous statement holds for arbitrary Kähler algebras.

Let now $\mathcal{J}$ and $\mathcal{H}$ be arbitrary normal Kähler algebras and let $h \rightarrow A(h)$ be a representation of the Lie algebra $\mathcal{H}$ by derivations of the algebra $\mathcal{J}$. We form the semi-direct sum

$$\mathcal{G} = \mathcal{J} + \mathcal{H}$$

of the algebras $\mathcal{J}$ and $\mathcal{H}$, so that for $h \in \mathcal{H}$, $x \in \mathcal{J}$

$$[h, x] = A(h)x \quad .$$

On $\mathcal{G}$ we define the operator j so that its restriction to $\mathcal{J}$ and $\mathcal{H}$ should coincide with the corresponding operators given there, and we define ϱ so that on $\mathcal{J}$ and $\mathcal{H}$ it coincides with the forms given there, and

$$\varrho(\mathcal{J}, \mathcal{H}) = 0 .$$

It is easy to see that in order that the algebra $\mathcal{G}$ constructed in this way should be a Kähler algebra (with $\mathcal{K} = 0$) it is necessary and sufficient that the following conditions be satisfied

(34) $$[A(jh) - j\,A(h),\ j] = 0 ,$$

(35) $$\varrho(A)x, y) + \varrho(x, A(h)\,y) = 0 ,$$

for all $h \in \mathcal{H}$, $x, y \in \mathcal{J}$.

For the normality of the algebra $\mathcal{G}$ it is necessary and sufficient that in addition the eigen-values of the operators $A(h)$, $h \in \mathcal{H}$, be real.

A representation A of a normal Kähler algebra $\mathcal{H}$ by real linear transformations of a hermitian vector space $\mathcal{J}$ will be called <u>symplectic</u>

if it satisfies condition (34) and (35), where we now take for j the complex structure operator in $\mathcal{J}$ and as ϱ the imaginary part of the hermitian scalar product.

The symplectic representation A is called normal if in addition the the eigen values of the operators $A(h)$, $h \in \mathcal{H}$ are real.

It is clear that the study of symplectic representation has to play an important role in the theory of Kähler algebras. We now find all normal symplectic representation of the two simplest normal Kähler algebras.

Lemma 3. Every normal simplectic representation A of a commutative Kähler algebra $\mathcal{H}$ is trivial, i.e. $A(h) = 0$ for $h \in \mathcal{H}$.

Proof. It is enough to prove the lemma for the two-dimensional algebra $\mathcal{H}$. In this case

$$\mathcal{H} = (js) + (s) ,$$

and $[js, s] = 0$. We set

$$p = A(js) , \quad q = A(s)$$

and represent the operators p and q in the form

$$p = p_1 + p_2 , \quad q = q_1 + q_2 ,$$

where p_1, q_1 are linear and p_2, q_2 anti-linear operators with respect to the complex structure of the space $\mathcal{J}$. The condition (34) for $h = s$ means that

$$p_2 = j\, q_2 \tag{36}$$

From the relation $[js, s] = 0$ it follows that

$$[p, q] = [p_1, q_1] + [p_2, q_2] + [p_1, q_2] + [p_2, q_1] = 0 .$$

Separating the real and imaginary parts of p, q we obtain :

$$[p_1, q_1] + [p_2, q_2] = 0 ,$$

$$[p_1, q_2] + [p_2, q_1] = 0 .$$

From the first of these equalities we obtain , subtracting (36),

$$[p_1, q_1] + 2 j q_2^2 = 0 ,$$

whence

(37) $$2 q_2^2 = j [p_1, q_1] = [jp_1, q_1] .$$

Condition (35) means that the linear part of the operator $A(h)$ is skew-symmetric and its anti-linear part is symmetric with respect to the real part of the hermitial scalar product on $\mathcal{J}$.

From (37) it follows that $S_p q_2^2 = 0$ and therefore $q_2 = 0$. So $q = q_1$ is a skew - symmetric operator. However by normality its eigen-values are real and therefore $q = 0$.

Analogously $p = 0$.

It is a little more difficult to prove the following

<u>Lemma</u> 4. Let $\mathcal{H}$ be and elementary normal Kähler algebra of dimension 2 i.e. $\mathcal{H} = (js) + (s)$, where $[js, s] = s$.

If A is a normal symplectic representation of the algebra $\mathcal{H}$ on the hermitian space $\mathcal{J}$, then the operator $p = A(js)$ is semi-simple and has eigen values $0, \pm \frac{1}{2}$.

We denote by $\mathcal{J}_\lambda$ the eigen spaces of p corresponding to the eigen value λ .

Then

$$j \mathcal{J}_\lambda = \mathcal{J}_{-\lambda}$$

$$A(s)x = \begin{cases} jx & \text{for } x \in \mathcal{J}_{-\frac{1}{2}} , \\ 0 & \text{for } x \in \mathcal{J}_{\frac{1}{2}} . \end{cases}$$

The proof of this lemma can be found in § 2 of [20] . Lemmas 3 and 4 will be used in the third part of these lectures.

§7. Structure of normal Kähler manifolds.

In the third part of these lectures we shall prove the "fundamental theorem" about the structure of normal Kähler algebras which is based on the following two assertions :

(*) Every non-degenerate normal Kähler algebra has a principal idempotent.

(**) Every normal Kähler algebra decomposes into a semi-direct sum of a commutative Kähler ideal and a non-degenerate normal Kähler sub-algebra.

Here we consider some consequences of this algebraic theorem.

Putting the assertion (*) together with proposition 2 we obtain the following theorem :

Theorem 1. Every homogeneous bounded domain admitting a transitive splittable solvable group of automorphisms is analytically isomorphic with some homogeneous Siegel domain of type II .

The assumption about the existence of a transitive splittable solvable group of automorphisms is essentially superfluous. This is shown in our article [25] .

From the assertion (**) we deduce the following theorem :

Theorem 2. Every normal Kähler manifold M admits a holomorphic fibering with the following properties :

1) the base is a homogeneous bounded domain ;

2) the fibre is locally flat (with respect to the induced Kähler structure) ;

3) the group of those automorphisms of M which preserve every fibre, acts transitively on the fibre.

Proof. Let G be an arbitrary transitive splittable solvable group of automorphisms of M , and let $\mathcal{G}$ be the corresponding normal Kähler algebra . We realize the Lie algebra $\mathcal{G}$ in the usual way as a Lie algebra of holomorphic vector fields on M .

We denote by $[\mathcal{G}]$ the complexification of $\mathcal{G}$ in the Lie algebra of all holomorphic vector fields on M, by [G] the simply connected complex Lie group which has $\mathcal{G}$ as its Lie algebra.

Furthermore let $[\mathcal{G}_0]$ be the sub-algebra of $\mathcal{G}$ formed by those vector fields which are zero at some point $p \in M$ and let $[G]_0$ be the corresponding connected sub-group of [G] .

The group [G] in general does not act on M , but M can be inbedded in the complex manifold $[G]/[G]_0$ as a domain so that the vector fields belonging to $[\mathcal{G}]$ become the corresponding vector fields on $[G] / [G]_0$.

Let now $\mathcal{G} = \mathcal{J} + \mathcal{H}$ be the decomposition of the Kähler algebra $\mathcal{G}$ into a semi-direct sum of a commutative Kähler ideal $\mathcal{J}$ and of a non-degenerate normal Kähler sub-algebra $\mathcal{H}$. We denote by I the connected normal sub-group of G corresponding to the ideal $\mathcal{J}$. The orbits of I are complex sub-manifolds of M . From the commutativity of I it follows that the Kähler structure induced on these sub-manifolds is locally flat.

The complexification $[\mathcal{J}]$ of the ideal $\mathcal{J}$ is an ideal in the Lie algebra $[\mathcal{G}]$.

We denote by [I] the corresponding connected normal sub-group of [G].

Since every orbit of I as a complex manifold is the quotient of a complex vector space by some lattice (see paragraph 2 of the introduction), the group of its analytic automorphisms is a complex Lie group. Consequently every vector field in $[\mathcal{J}]$ determines a one parameter group of mappings of the manifold M. This means that M is contained in $[G]/[G]_0$ as a domain invariant under the action of $[I]$.

The fibering of M by the orbits of I is a restriction of the holomorphic fibering

$$[G]/[G]_0 \longrightarrow [G]/[I].[G]_0$$

and therefore is itself holomorphic.

From the preceeding if follows that this fibering satisfies conditions 2) and 3) of the theorem. Its base is the complex manifold M/I, the complex structure of which is determined by the operator j on $\mathcal{G}/\mathcal{J}$. We show now that the group G decomposes into the semi-direct product of the normal sub-group I and of the sub-group H corresponding to the sub-algebra $\mathcal{H} \subset \mathcal{G}$.

Let $\tilde{G}$ be the universal covering of G. It is obvious that it decomposes into a semi-direct product of the normal sub-group $\tilde{I}$ corresponding to the ideal $\mathcal{J}$ and the sub-group $\tilde{H}$ corresponding to $\mathcal{H}$. Like G, the group $\tilde{G}$ is also a splittable solvable Lie group and therefore its center is connected.

It is not difficult to prove that the center of $\mathcal{G}$ is in $\mathcal{J}$ (see Lemma 6, §1 of part 3). Consequently the kernel of the natural holomorphic $\tilde{G} \longrightarrow G$ is in $\tilde{I}$. This means that the group G decomposes into the semi-direct product of I and H, and the group H is simply connected. We see now that H acts transitively and without fixed points on M/I. Under the natural identification of the tangent space of M/I with the Lie algebra $\mathcal{H}$, the complex structure on the tangent space is given by the operator j on $\mathcal{H}$. Therefore M/I

is exactly the h.b.d. corresponding to the non-degenerate normal Kähler algebra $\mathcal{H}$. This proves statement 1) of the theorem.

Theorem 2 is a special case of the "fundamental hypothesis" formulated in § 5 of the introduction.

PART III - Structure of normal Kähler algebras.

§ 1. Statement of the fundamental theorem and its corollaries.

As it was shown in part II of these lectures, the study of normal Kähler manifolds reduces to the study of normal Kähler algebras (cf. §4, 7).

The structure of normal Kähler algebras is described by the following theorem :

Fundamental theorem : Every normal Kähler algebra can be decomposed into a semi-direct sum

$$\mathcal{G} = \mathcal{J} + \mathcal{H} \tag{1}$$

where $\mathcal{J}$ is a commutative Kähler ideal and $\mathcal{H}$ is a non-degenerate Kähler sub-algebra. Every-non-degenerate normal Kähler algebra has an element s with the following properties :

1) $[js, s] = s$;

2) The operator $\mathrm{ad}_{\mathcal{H}}\, js$ is semi-simple and has eigenvalues $0, 1, 1/2$.

3) If $\mathcal{H}_\lambda$ denotes the eigen space of $\mathrm{ad}_{\mathcal{H}}\, js$ corresponding

to the eigen-value λ , then $j\mathcal{H}_\lambda = \mathcal{H}_{1-\lambda}$

We also note that $[\mathcal{H}_\lambda, \mathcal{H}_\mu] \subset \mathcal{H}_{\lambda+\mu}$.

The elements is called the principal idempotent of the Kähler algebra $\mathcal{G}$.

The outline of the proof of this theorem will be the following: we show that $\mathcal{G}$ admits a decomposition (1) where $\mathcal{J}$ is a commutative Kähler ideal and $\mathcal{H}$ is a Kähler sub-algebra possessing an element s with properties 1), 2), 3) .

In order to obtain from here the second statement of the theorem it is enough to show that in a non-degenerate normal Kähler algebra there are no non-zero commutative Kähler sub-algebras and this follows from the fact that in such an algebra $[jx, x] = 0$ (see § 4 of part II) .

The first statement of the theorem will follows from the following lemma.

Lemma 1 : If in a normal Kähler algebra $\mathcal{H}$ there exists an element s having the properties listed in the theorem, then the algebra $\mathcal{H}$ is non-degenerate.

Proof. Transcribing the condition of the form ϱ being closed for $x \in \mathcal{H}_\lambda$, $y \in \mathcal{H}_\mu$, $z = js$ we obtain

(2) $$(\lambda + \mu)\, \varrho(x,y) = (js, [x,y]) .$$

From this equality it follows that $\varrho(x,y) = 0$ for $x, y \in \mathcal{H}_1$ and therefore (by the condition of hermiticity (K A 5)) also for $x, y \in \mathcal{H}_0$.

It is clear now that for every $x, y \in \mathcal{H}$

$$\varrho(x,y) = \omega([x,y]) ,$$

where

$$\omega(u) = \begin{cases} \frac{1}{\lambda} \quad \varrho(js, u) & \text{for } u \in \mathcal{H}_\lambda, \ \lambda \neq 0, \\ 0 & \text{for } u \in \mathcal{H}_0 . \end{cases}$$

2. We indicate some consequences of the properties of the decomposition (1) listed in the theorem.

We note first that

(3) $$[s, x] = j\,x \quad \text{for all} \quad x \in \mathcal{H}_0 .$$

This follows immediately from the integrability condition (K A 3) .

The other consequences we formulate as lemmas (of which only lemma 3 will not be used in the sequal) .

Lemma 2 : The operator $ad_I\, js$ is semi-simple and has eigenvalues $0, \pm\frac{1}{2}$, and

(4) $$j\,\mathcal{J}_\lambda = \mathcal{J}_{-\lambda} ,$$

(5) $$[s, x] = jx \quad \text{for} \quad x \in \mathcal{J}_{-\frac{1}{2}} .$$

Proof. We consider the representation of the two-dimensional Kähler algebra generated by s and js on the ideal $\mathcal{J}$ induced by the adjoint representation of the algebra $\mathcal{G}$. It is easy to see that it is symplectic in the sense of the definition given in the second part.

Therefore lemma 2 is an immediate consequence of the lemma on symplectic representation

Lemma 3 : The sub-spaces $\mathcal{H}_\lambda$ and $\mathcal{J}_\mu$ are mutually orthogonal with respect to the canonical scalar product on $\mathcal{G}$.

Proof. We set $\mathcal{G}_\lambda = \mathcal{J}_\lambda + \mathcal{H}_\lambda$. The formula (2) is clearly true all $x \in \mathcal{G}_\lambda$, $y \in \mathcal{G}_\mu$. It shows in particular that if $[x, y] = 0$

and $\lambda + \mu \neq 0$, then $\varrho(x,y) = 0$.

Now we distinguish three cases

a) Let $u \in \mathcal{H}_\varrho$, $v \in \mathcal{H}_\sigma$ $\varrho < \sigma$. Then $ju \in \mathcal{H}_{1-\varrho}$ (condition 3) of the theorem) and $[ju, v] = 0$ since $1 - \varrho + \sigma > 1$ and is not and eigenvalues' of ad js. Applying (2) to $x = ju$, $y = v$ we see that $(u, v) = \varrho(x,y) = 0$.

b) Let $u \in \mathcal{J}_\varrho$, $v \in \mathcal{J}_\sigma$, $\varrho < \sigma$. Then by (4) ju $\mathcal{J}_{-\varrho}$ and from the commutativity of $\mathcal{J}$ it follows that $[ju, v] = 0$.

Since $-\varrho + \sigma > 0$ applying (2) to $x = ju$, $y = v$, we obtain that $(u, v) = 0$.

c) It remains to show that $\varrho(\mathcal{J}, \mathcal{H}) = 0$. We prove first that $\varrho(js, \mathcal{J}_{\frac{1}{2}}) = 0$. Any element of $\mathcal{J}_{\frac{1}{2}}$ can be represented in the form jw where $w \in \mathcal{J}_{-\frac{1}{2}}$. We have

$$\varrho(js, jw) = \varrho(s, w),$$

and on the other hand, by (5)

$$\varrho(js, jw) = \varrho(js, [s, w]) =$$

$$= \varrho([js, s], w) + \varrho(s, [js, w]) = \frac{1}{2}\varrho(s, w),$$

consequently $\varrho(js, jw) = 0$.

Let now $u \in \mathcal{J}_\varrho$, $v \in \mathcal{H}_\sigma$ and $\varrho < \sigma$. Then $ju \in \mathcal{J}_{-\varrho}$ and $[ju, v] \in \mathcal{J}_{\frac{1}{2}}$ since $-\varrho + \sigma > 0$.

Applying (2) to $x = ju, y = v$ and considering that $\varrho(js, \mathcal{J}_{\frac{1}{2}}) = 0$ we find that $(u, v) = 0$. If $\varrho \geq \sigma$, then $-\varrho < 1-\sigma$ and

$$(\mathcal{J}_\rho , \mathcal{H}_\sigma) = (j\mathcal{J}_\rho , j\mathcal{H}_\sigma) = (\mathcal{J}_{-\rho}, \mathcal{H}_{1-\sigma}) = 0 .$$

Lemma 4. $[\mathcal{H}_0 , \mathcal{J}_0] = 0$.

Proof. We consider the operator $A = \mathrm{ad}_{\mathcal{J}_0} h$ where $h \in \mathcal{H}_0$. Since $[\mathcal{H}_1, \mathcal{J}_0] = 0$, using the integrability condition it follows that A commutes with j. By the closure of the form ρ and the commutativity of $\mathcal{J}$,

$$\rho(Ax, y) + \rho(x, Ay) = 0 \qquad (x, y \in \mathcal{J}_0) .$$

This shows that the operator A is skew-symmetric with respect to the canonical scalar product. Since its eigenvalues must be real we have $A = 0$.

Lemma 5; Let $\lambda + \mu + \nu = 0$.
If the element $h \in \mathcal{H}_\lambda$ commutes with $\mathcal{J}_\mu$ then it commutes also with $\mathcal{J}_\gamma$.

Proof. Let $a \in \mathcal{J}_\gamma$. Then by the closure of the form ρ and by the condition of the lemma, $\rho([h, a], x) = 0$ for all $x \in \mathcal{J}_\mu$. Since, by lemma 2, $j[h, a]$, we have $\rho([h, a], j[h, a]) = 0$ and therefore $h, a = 0$.

Lemma 6. The ideal $\mathcal{J}$ coincides with the set of all $g \in \mathcal{G}$ for which $[jg, g] = 0$.

Proof. It is enough to prove that if $[jg, g] = 0$ then $g \in \mathcal{J}$. Let $g = x + h$ where $x \in \mathcal{J}$, $h \in \mathcal{J}$. Then

$$0 = [jg, g] = [jh, h] \pmod{\mathcal{J}} .$$

Since $\mathcal{H}$ is a non-degenerate Kähler algebra, we have $h = 0$, which was to be shown.

Lemma 7. The center $\mathcal{Z}$ of the algebra $\mathcal{G}$ is contained in $\mathcal{J}_0$ and is j - invariant.

Proof. Lemma 6 shows that $\mathcal{Z} \supset \mathcal{J}$. Furthermore it is clear that $\mathcal{Z} \subset \mathcal{J}_0$.

From lemma 4 and the fact that $[\mathcal{H}_1, \mathcal{J}_0] = 0$, it follows that $\mathcal{Z}$ coincides with the centralizer of $\mathcal{H}_{\frac{1}{2}}$ in $\mathcal{J}_0$. Let $a \in \mathcal{Z}$. From the integrability condition we find that

$$[ja, jh] = j\,[ja, h] \qquad \text{for all} \quad h \in \mathcal{H}_{\frac{1}{2}}$$

so that

$$\rho(j\,[ja, h], [ja, h]) = \rho([ja, jh], [ja, h]).$$

Using the condition (K A 7) of the closure of ρ and taking into account that $\big[ja, [ja, h]\big] = 0$ and $\big[jh, [ja, h]\big] = 0$ we obtain

$$(j\,[ja, h], [ja, h]) = 0 .$$

This proves that $[ja, h] = 0$ and $ja \in \mathcal{Z}$.

The purpose of the remaining paragraphs of this part is the proof of the fundamental theorem. Here is the logical scheme of this proof.

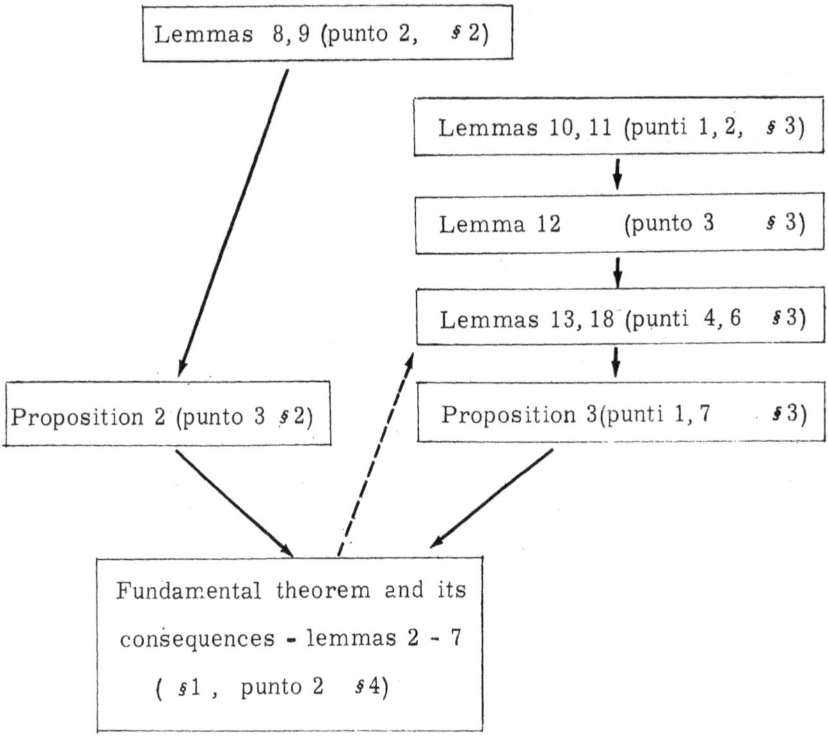

The theorem will be proved by induction with respect to the dimension of the Kähler algebra. The dotted arrow indicates that in the proof of lemmas 13 - 18 we use the induction hypothesis .

If it is known in advance that the algebra is non-degenerate, then the proof becomes considerably simpler. In this case the whole right-hand side of the scheme (§ 3 in the text) turns out to be unnecessary and so do lemmas 2 - 7 from § 1 .

2. Selection of an elementary Kähler ideal.

1. The fundamental theorem will be proved by induction on the dimension of the algebra $\mathcal{G}$ together with the following assertion :

Proposition 1. In every normal Kähleralgebra there is either an elementary or a commutative non-zero Kähler ideal.

Assuming that the fundamental theorem holds for algebras of dimension less than n, in §§ 2 and 3 we prove proposition 1, and in paragraph 4 we prove the theorem for algebras of dimension n.

§ 2. Since the Lie algebra $\mathcal{G}$ is solvable and splittable, there is a one-dimensional ideal in it. In this ideal we choose a non-zero element r so that $[jr, r] = r$ or $[jr, r] = 0$. In this paragraph we show that in the first case the element r is contained in some elementary Kähler ideal, and in § 3 we show that in the second case it is contained in a commutative Kähler ideal.

To start we prove two lemmas connected with these two situations.

Lemma 8. Let $\mathcal{P}$ be the sub-space consisting of those elements $g \in \mathcal{G}$ for which

(6) $$[g, r] = [jg, r] = 0 .$$

Then $\mathcal{P}$ is invariant under j and $\operatorname{ad} jr$; moreover the operator $\operatorname{ad}_{\mathcal{P}} jr$ commutes with j.

Proof. The invariance of $\mathcal{P}$ with respect to j is immediate from the definitions. From the integrability condition we have

(7) $$[jr, jp] = j\,[jr, p]$$

for all $p \in \mathcal{P}$. Using the Jacobi identity and (7) we find

$$\big[[jr, p], r\big] = 0 ,$$

$$\big[j\,[jr, p], r\big] = \big[[jr, jp], r\big] = 0 ,$$

so that $[jr, p] \in \mathcal{P}$. From (7) the commutativity of $\operatorname{ad} jr$ and j follows.

Lemma 9. For all $u, v \in \mathcal{G}$

(8) $$\frac{d}{dt}\,\rho(e^{tadjr}u, e^{tadjr}u, e^{tadjr}v) = \rho(jr, e^{tadjr}[u,v]) .$$

Proof.

$$\frac{d}{dt}\,(e^{tadjr}u,\ e^{tadjr}v) =$$

$$= \rho([jr, e^{tadjr}u], e^{tadjr}v) + \rho(e^{tadjr}u, [jr, e^{tadjr}v]) =$$

$$= \rho(jr, [e^{tadjr}u, e^{tadjr}v]) = \rho(jr, e^{tadjr}[u,v]) .$$

(In the second step we used the fact that the form ρ is closed).

3. In this point we prove proposition 2 from which proposition 1 follows in the case where $[jr, r] = r$.

For this we will not use the induction hypothesis concerning the fundamental theorem.

Proposition 2. If in the normal Kähler algebra there is a one-dimensional ideal (r)(1) such that $[jr, r] = r$, then the algebra $\mathcal{G}$ can be decomposed into a direct sum of sub-spaces

(9) $$\mathcal{G} = (jr) + (r) + \mathcal{U} + \mathcal{G}' ,$$

where

1) $\mathcal{N} = (jr) + (r) + \mathcal{U}$ is an elementary Kähler ideal and $[jr, u] = \frac{1}{2}u$ for $u \in \mathcal{U}$;

2) $\mathcal{G}'$ is a Kähler sub-algebra orthogonal to $\mathcal{N}$ and commuting with jr and r.

(1) By (r) we denote the one-dimensional ideal spanned by the vector r.

Proof. Let $\mathcal{P}$ be the sub-space constructed in lemma 8. By the relation $[jr, r] = r$ one can find for any element $g \in \mathcal{G}$ numbers a and b in a unique way, so that

$$g - ajr - br \in \mathcal{P}.$$

This means that $\mathcal{G}$ decomposes into a direct sum of sub-spaces

$$\mathcal{G} = (jr) + (r) + \mathcal{P}. \tag{10}$$

Using lemma 9 we shall study the eigenvalues of ad jr on $\mathcal{P}$.

a) If in (8) we set $u = r$, $v \in \mathcal{P}$ then the right hand side is 0. Using that

$$e^{t\,\mathrm{ad}\,jr}\, r = e^{t} r \tag{11}$$

we obtain

$$\rho(r, e^{t\,\mathrm{ad}\,jr} v) = a\, e^{-t} \quad \text{for all } v \in \mathcal{P} \tag{12}$$

According to lemma 8 the operator ad j r commutes with j on $\mathcal{P}$. Therefore

$$\rho(jr, e^{t\,\mathrm{ad}\,jr} v) = a\, e^{-t} \quad \text{for all } v \in \mathcal{P}. \tag{13}$$

Finally for any $g \in \mathcal{G}$ by (10) and (11) we obtain

$$\rho(jr, e^{t\,\mathrm{ad}\,jr} g) = a\, e^{-t} + b\, e^{t} . \tag{14}$$

Formula (8) now gives for all $u, v \in \mathcal{G}$

$$\rho(e^{t\,\mathrm{ad}\,jr} u, e^{t\,\mathrm{ad}\,jr} v) = a\, e^{-t} + b e^{t} + c . \tag{15}$$

If $u \in \mathcal{P}$, then in this equality we can change over to a scalar product using that $\mathrm{ad}_{\mathcal{P}}\, jr$ and j commute. We obtain

$$(e^{t\,\mathrm{ad}_{\mathcal{P}} r} u, e^{t\,\mathrm{ad}\,jr} v) = a e^{-t} + b e^{t} + c \quad \text{for all } p \in \mathcal{P},\ v \in \mathcal{G}. \tag{16}$$

b) Let $p \in \mathcal{P}$ be an eigenvector for ad jr corresponding to eigen-value λ . Then $e^{tadjr} p = e^{\lambda t} p$.

By formula (16)

$$e^{2\lambda(t)}(p,p) = ae^{-b} + be^{t} + c \ .$$

Since $(p,p) \neq 0$, λ can have only the values $0, \pm\frac{1}{2}$. We check that the operator ad jr is semi-simple on $\mathcal{P}$ (and therefore also on $\mathcal{G}$). Let p and q be vectors in $\mathcal{P}$ such that $[jr, p] = \lambda p$, $[jr, z] = \lambda q + p$. Then $e^{tadjr} q = e^{\lambda t} q + te^{\lambda t} p$ and by formula (16)

$$(e^{tadjr} p, e^{tadjr} q) = e^{2\lambda t}(p,q) + (te^{2\lambda t} (p,p) =$$

$$= a e^{-t} + b e^{t} + c \ .$$

Being $(p,p) = 0$, this equality is impossible.

c) Since the operator $ad_{\mathcal{P}} jr$ has no eigen-value -1 , we have $a = 0$ in (12) . Looking at the formulas that were deduced from (12), we see that in each of them $a = 0$, and therefore $-\frac{1}{2}$ is not an eigen-value of $ad_{\mathcal{P}}$ jr .

We denote by $\mathcal{P}_\lambda$ the eigen-space of $ad_{\mathcal{P}}$ jr corresponding to the eigen-value λ . The algebra $\mathcal{G}$ decomposes into a direct sum of subspaces

$$(17) \qquad \mathcal{G} = (jr) + (r) + \mathcal{P}_{\frac{1}{2}} + \mathcal{P}_0 \ .$$

Since $ad_{\mathcal{P}}$ jr commutes with j the subspaces $\mathcal{P}_\lambda$ are invariant under j . Formulas (12) , (13) and (16) show that $\mathcal{P}_0$ is orthogonal to $(jr) + (r) + \mathcal{P}_{\frac{1}{2}}$.

Since $\mathcal{P}_0 \subset \mathcal{P}$ we have $[r, \mathcal{P}_0] = 0$. Consideration of the eigen-values shows that $N = (jr) + (r) + \mathcal{P}_{\frac{1}{2}}$ is an ideal.

The sub-space $\mathcal{P}_0$, being the orthogonal complement of N, is a sub-algebra (see § 6, part 2) setting $\mathcal{U} = \mathcal{P}_{\frac{1}{2}}$ $\mathcal{G}' = \mathcal{P}_0$, we obtain the decomposition (9), which has the required properties.

§ 3. Commutative Kähler ideal.

1. In this paragraph we consider the case where $[jr, r] = 0$. Under the hypothesis that the fundamental theorem holds for algebras of lower dimension then $\mathcal{G}$ we prove

Proposition 3 : If the normal Kähler algebra $\mathcal{G}$ contains a one-dimensional ideal (r) such that $[jr, r] = 0$, then this ideal is contained in a commutative Kähler ideal $\mathcal{N}$.

In all the lemmas of this paragraph the hypothesis of proposition 3 is assumed. By $\mathcal{P}$ we denote the sub-space constructed in lemma 8.

Lemma 10 : $[jr, \mathcal{G}] \subset \mathcal{P}$.

Proof. For any $g \in \mathcal{G}$ from the Jacobi identity and from the integrability condition we have

$$\big[[jr, g], r\big] = \big[jr, [g, r]\big] = 0 ,$$

$$\big[j\,[jr, g], r\big] = \big[[jr, jg], r\big] = 0 ,$$

i.e. $[jr, g] \in \mathcal{P}$.

2. Lemma 11 : $(\text{ad}\ jr)^2 = 0$.

Proof: a) We show that the sub-space $[jr, \mathcal{P}]$ is orthogonal to $(jr) + (r)$. Let $p \in \mathcal{P}$. By the closure of the form ρ we have

$$\varrho(r, [jr, p]) = \varrho(jr, [r, p]) = 0 .$$

Using lemma 8 we also obtain

$$\varrho(jr, [jr, p]) = -\varrho(jr, j[jr, jp]) = -\varrho(r, [jr, jp]) = 0 .$$

b) From lemma 10 and a) it follows that

(18) $$\varrho(jr, (\mathrm{ad}\, jr)^2 \mathfrak{g}) = 0 .$$

Formula (8) shows then that for any $u, v \in \mathfrak{g}$

$$\frac{d^3}{dt^3} \varrho(e^{t\,\mathrm{ad}\, jr} u, e^{t\,\mathrm{ad}\, jr} v) = \frac{d^2}{dt^2} \varrho(jr, e^{t\,\mathrm{ad}\, jr} [u, v]) =$$

$$= \varrho(jr, [jr, [jr, e^{t\,\mathrm{ad}\, jr} [u, v]]]) = 0 ,$$

i.e.

(19) $$(e^{t\,\mathrm{ad}\, jr} u, e^{t\,\mathrm{ad}\, jr} v) = at^2 + bt + c .$$

Whence for $p \in \mathcal{P}$

(20) $$(e^{t\,\mathrm{ad}\, jr} p, e^{t\,\mathrm{ad}\, jr} v) = at^2 + bt + c .$$

Setting $v = p$ we obtain that the operator $\mathrm{ad}\, jr$ has no non-zero eigenvalues on $\mathcal{P}$ (and therefore on $\mathfrak{g}$).

c) We assume that $(\mathrm{ad}\, jr)^2 \neq 0$. Then there exist elements $x, y, z \in \mathfrak{g}$ such that:

$$[jr, x] = 0 , \quad [jr, y] = 0 , \quad [jr, z] = 0 .$$

It is clear that

$$e^{t\,\mathrm{ad}\, jr} x = x , \quad e^{t\,\mathrm{ad}\, jr} y = y + tx, \quad e^{t\,\mathrm{ad}\, jr} z = z + ty + \frac{t^2}{2} x .$$

By lemma 10 $y \in \mathcal{P}$ and we can substitute $p = y$, $v = z$ in (20). We get

$$(y+tx, z+ty+\frac{t^2}{2}x) = at^2 + bt + c \quad .$$

Since $(x,x) \neq 0$, this is impossible.

3. With the aid of the operator ad jr we construct a j-invariant filtration of the Lie algebra $\mathfrak{g}$. Let $\mathfrak{U} = \mathfrak{g}/(jr) + (r)$ and let π be the natural projection of $\mathfrak{g}$ onto $\mathfrak{U}$.

We denote by A the operator induced by adjr on $\mathfrak{U}$. From the integrability condition it follows that A commutes with j on $\mathfrak{U}$. We set

$$\mathfrak{g}^{(-1)} = \mathfrak{g} ,$$

$$\mathfrak{g}^{(o)} = \pi^{-1}(\text{Ker } A) ,$$

$$\mathfrak{g}^{(1)} = \pi^{-1}(\text{Im } A) ,$$

$$\mathfrak{g}^{(2)} = (jr) + (r) \quad .$$

<u>Lemma</u> 12 : The sub-spaces $\mathfrak{g}^{(i)}$ form a j-invariant filtration of the Lie algebra $\mathfrak{g}$. Furthermore

(21) $$[\mathfrak{g}^{(1)}, \mathfrak{g}^{(1)}] = 0 \quad .$$

<u>Proof:</u> a) From lemma 11 it follows that $A^2 = 0$. Therefore Ker A $\supset$ Im A , and we have the inclusions

$$\mathfrak{g}^{(-1)} \supset \mathfrak{g}^{(o)} \supset \mathfrak{g}^{(1)} \supset \mathfrak{g}^{(2)} \quad .$$

The invariance of the sub-spaces $\mathfrak{g}^{(i)}$ with respect to j follows from the commutativity of A with j .

b) We prove the commutativity of $\mathfrak{g}^{(1)}$. Let $g_1, g_2 \in \mathfrak{g}$ and $u_1 = [jr, g_1]$, $u_2 = [jr, g_2]$. Since $(\text{adjr})^2 = 0$,

$$0 = (\mathrm{ad}\,jr)^2\,[g_1, g_2] = 2\,[u_1, u_2] \;.$$

Therefore

$$[[jr, \mathfrak{g}]\ , [jr, \mathfrak{g}]] = 0 \;.$$

Since

$$\mathfrak{g}^{(1)} = [jr, \mathfrak{g}] + (jr) + (r) \;,$$

it remains to prove that

$$[(jr) + (r)\ , [jr, \mathfrak{g}]] = 0$$

but this follows from lemmas 10 and and 11.

c) It follows immediately from the definitions that

$$[\mathfrak{g}^{(-1)}\ , \mathfrak{g}^{(2)}] \subset \mathfrak{g}^{(1)}\ , [\mathfrak{g}^{(0)}\ , \mathfrak{g}^{(2)}] \subset \mathfrak{g}^{(2)} \;.$$

d) We show that

$$[\mathfrak{g}^{(0)}\ , \mathfrak{g}^{(1)}] \subset \mathfrak{g}^{(1)} \;.$$

Let $k \in \mathfrak{g}^{(0)}$, $u \in \mathfrak{g}^{(1)}$. Since $[\mathfrak{g}^{(0)}\ , \mathfrak{g}^{(2)}] \subset \mathfrak{g}^{(1)}$, it is enough to consider the case where $u = jr, g$ and $g \in \mathfrak{g}$. We have

$$[jr, [k, g]] = [k, u] + [[jr, k]\ , g] \;.$$

Since

$$[jr, [k, g]]\ , [[jr, k]\ , g] \in \mathfrak{g}^{(1)} \;,$$

we also have

$$[k, u] \in \mathfrak{g}^{(1)} .$$

e) We prove that

$$[\mathfrak{g}^{(-1)}\ , \mathfrak{g}^{(1)}] \subset \mathfrak{g}^{(0)} \;.$$

Let $g \in \mathfrak{g}$, $u \in \mathfrak{g}^{(1)}$. Since $[jr, u] = 0$ and $[jr, g] \in \mathfrak{g}^{(1)}$, we obtain using the Jacobi identity and the already proved commutativity of $\mathcal{G}^{(1)}$ that $[jr, [g, u]] = 0$. Therefore $[g, u] \in \mathcal{G}^{(0)}$.

f) It remains to prove that $\mathcal{G}^{(0)}$ is a sub-algebra. Let $g_1, g_2 \in \mathcal{G}^{(0)}$. Since $[\mathcal{G}^{(2)}, \mathcal{G}^{(0)}] \subset \mathcal{G}^{(2)}$ we have

$$[jr, [g_1, g_2]] = [[jr, g_1], g_2] + [g_1, [jr, g_2]] \in \mathcal{G}^{(2)} .$$

Consequently $[g_1, g_2] \in \mathcal{G}^{(0)}$.

The proof of lemma 12 is finished.

4. If in the filtration constructed above $\mathcal{G}^{(0)} = \mathcal{G}$ then $\mathcal{N} = \mathcal{G}^{(2)}$ is a commutative Kahler ideal in $\mathcal{G}$ and proposition 3 is proved. Therefore in the following we shall assume that $\mathcal{G}^{(0)} \neq \mathcal{G}$.

The induction hypothesis may be applied to the Kähler algebra $\mathcal{G}^{(0)}$. Let

$$\mathcal{G}^{(0)} = \mathcal{J} + \mathcal{K} \tag{22}$$

be the decomposition corresponding to this theorem.

Lemma 13 : $\mathcal{G}^{(1)} \subset \mathcal{J}$.

Proof : Let $g \in \mathcal{G}^{(1)}$. Then $jg \in \mathcal{G}^{(1)}$. From (21) it follows that $[jg, g] = 0$. By lemma 6 this means that $g \in \mathcal{J}$.

Lemma 14 : $[\mathcal{G}, \mathcal{J}] \subset \mathcal{G}^{(0)}$.

Proof : Let $g \in \mathcal{G}$, $x \in \mathcal{J}$. Then $[jr, g] \in \mathcal{G}^{(1)}$ and by lemma 14 $[[jr, g], x] = 0$. Therefore

$$[jr, [g, x]] = [g, [jr, x]] = 0 .$$

Thereby the lemma is proved.

From lemma 14 it follows in particular that if the sub-algebra $\mathcal{G}^{(0)}$ is commutative, then it is an ideal in $\mathcal{G}$, so that in this case proposition 3 is proved.

In the following we assume that $\mathfrak{g}^{(0)}$ is not commutative, i.e. $\mathcal{H} = 0$.

We denote by s the principal Idem potent of H . We denote by $\mathfrak{g}_\lambda$ the subspace formed by the vectors which are annihilated by some power of the operator ad js - λ . It is clear that

$$\mathfrak{g} = \Sigma\, \mathfrak{g}_\lambda \quad , \quad [\mathfrak{g}_\lambda, \mathfrak{g}_\mu] \subset \mathfrak{g}_{\lambda+\mu}, \ \mathfrak{g}_\lambda \cap \mathfrak{g}^{(0)} = \mathfrak{g}^{(0)}_\lambda .$$

<u>Lemma</u> 15 : If $\lambda + \mu > 0$ or $\lambda = \mu = 0$, then

$$\left[[\mathfrak{g}, \mathcal{J}_\lambda], \mathcal{J}_\mu\right] = 0 .$$

<u>Proof.</u> Let $g \in \mathfrak{g}_\nu$, $x \in \mathcal{J}_\lambda$. Then by lemma 14

$$[g, x] \in \mathfrak{g}^{(0)}_{\lambda+\mu} \subset \mathcal{H}_{\lambda+\mu} + \mathcal{J} .$$

According to lemma 5 it is sufficient to prove that $\left[[g, x], \mathcal{J}_{-(\lambda+\mu+\nu)}\right] = 0$.

Let $y \in \mathcal{J}_{-(\lambda+\mu+\nu)}$. By the commutativity of $\mathcal{J}$

$$\left[[g, x] \ , y\right] = \left[[g, y] \ , x\right] = 0 .$$

We have

$$[g, y] \in \mathfrak{g}^{(0)}_{-(\lambda+\mu)} \subset \mathcal{H}_{-(\ +\ )} + \mathcal{J} .$$

If $\lambda + \mu > 0$ then $[g, y] \in \mathcal{J}$ since the operator ad$_{\mathcal{H}}$js has only non-negative eigen-values. Consequently in this case $\left[[g, y], x\right] = 0$. If $\lambda = \mu = 0$ then $[g, y] \in \mathcal{H}_0 + \mathcal{J}$, $x \in \mathcal{J}_0$ and $\left[[g, y] \ , x\right] = 0$ by lemma 4.

5. We consider the graded Lie algebra

$$(23) \qquad \bar{\mathcal{G}} = \bar{\mathcal{G}}^{(-1)} + \bar{\mathcal{G}}^{(0)} + \bar{\mathcal{G}}^{(1)} + \bar{\mathcal{G}}^{(2)} ,$$

which is associated to the filtered Lie algebra $\mathcal{G}$. For every element $g \in \mathcal{G}^{(i)}$ we denote by $\bar{g}$ the corresponding element of $\bar{\mathcal{G}}^{(i)}$ (1).

From (21) it follows that

$$(24) \qquad \left[\bar{\mathcal{G}}^{(1)} , \bar{\mathcal{G}}^{(1)}\right] = 0 .$$

We define on $\bar{\mathcal{G}}^{(-1)}$ a trilinear operation $(a, b, c) \longrightarrow a\,b\,c$ by the formula

$$(25) \qquad a\,b\,c = \left[\left[\left[\bar{jr}, a\right], b\right], c\right] .$$

We established some properties of this operation.

<u>Lemma</u> 16: 1) The operation (25) is commutative.

2) $$\left[\bar{jr}, abc\right] = \left[\left[\left[\bar{jr}, a\right], b\right], \left[\bar{jr}, c\right]\right] .$$

<u>Proof:</u> 1) By the Jacobi identity

$$\left[\left[\left[\bar{jr}, a\right], b\right], c\right] - \left[\left[\left[\bar{jr}, b\right], a\right], c\right] = $$
$$= \left[\left[\bar{jr}, [a,b]\right], c\right] = 0$$

since $\left[\bar{\mathcal{G}}^{(-1)} , \bar{\mathcal{G}}^{(-1)}\right] = 0$. This shows that abc = bac. One proves analogously that abc = acb.

2) We take the commutator of both sides of equation (25) with jr and use the Jacobi identity. From the properties of the graduation and from (24) it follows that $\left[\bar{jr}, \left[\bar{jr}, a\right]\right] = 0$, $\left[\left[\bar{jr}, a\right], \left[\bar{jr}, b\right]\right] = 0$, so that, of the three terms on the right hand side, there remains only $\left[\left[\left[\bar{jr}, a\right], b\right], \left[\bar{jr}, c\right]\right]$ which we have to prove

(1) If we regard the same element $g \in \mathcal{G}^{(i)}$ as an element of $\mathcal{G}^{(i-1)}$ then $\bar{g} = 0$. However in the following it will always be clear which of the subspaces $\mathcal{G}^{(i)}$ we have in mind.

The following lemma is fundamental for this paragraph .

Lemma 17 : abc = 0 for all $a, b, c \in \overline{\mathcal{G}}^{(-1)}$.

Proof. To the decomposition (22) of the algebra $\overline{\mathcal{G}}^{(o)}$, there corresponds the decomposition

(26) $$\overline{\mathcal{G}}^{(o)} = \overline{\mathcal{J}} + \overline{\mathcal{K}}$$

of the algebra $\overline{\mathcal{G}}^{(o)}$. By lemma 13

(27) $$\left[\overline{\mathcal{G}}^{(1)} + \overline{\mathcal{G}}^{(2)} , \overline{\mathcal{J}}\right] = 0 ,$$

and from lemma 14 if follows that

(28) $$\left[\overline{\mathcal{G}}^{(-1)} , \overline{\mathcal{J}}\right] = 0 .$$

From lemma 13 if follows that $r \in \mathcal{J}$. Since (r) is an ideal $[js, r] = \alpha r$. By lemma 2 $\alpha = 0$ or $\pm \frac{1}{2}$, and if $\alpha = -\frac{1}{2}$ then $[s, r] = jr \notin (r)$, so that this case is impossible.

Furthermore, if $r \in \mathcal{J}_\alpha$, then $jr \in \mathcal{J}_{-\alpha}$. Going over to the algebra $\overline{\mathcal{G}}$ we obtain the following relation

(29) $$\left[\overline{js}, \overline{jr}\right] = -\alpha jr \quad \text{where} \quad \alpha = 0 \text{ or } \frac{1}{2} .$$

From the definition of the sub-spaces $\mathcal{G}^{(i)}$ it is clear that the operator ad $\overline{jr}$ maps $\overline{\mathcal{G}}^{(-1)}$ isomorphically onto $\overline{\mathcal{G}}^{(1)}$. The operator ad js is semi-simple on $\overline{\mathcal{G}}^{(1)}$ and has eigenvalues 0 , $\pm \frac{1}{2}$ on it.

Relation (29) shows that it is semi-simple also on $\overline{\mathcal{G}}^{(-1)}$ and has eigenvalues α , $\pm \frac{1}{2} + \alpha$ there .

Let $a \in \overline{\mathcal{G}}_\lambda^{(-1)}$, $b \in \overline{\mathcal{G}}_\mu^{(-1)}$, $c \in \mathcal{G}_\nu^{(-1)}$, so that

(30) $$abc = \left[\left[[\overline{jr}, a] , b\right] c\right] \neq 0 .$$

Then also

(31) $$[\bar{jr}, abc] = [[[\bar{jr}, a], b], [\bar{jr}, c]] \neq 0 .$$

From what it has been said above it is clear that

(32) $$\lambda, \mu, \nu = \alpha \text{ or } \pm \frac{1}{2} + \alpha .$$

Furthermore

$$[[jr, a], b] \in \bar{\mathcal{G}}^{(o)}_{\lambda+\mu-\alpha} = \bar{\mathcal{K}}_{\lambda+\mu-\alpha} + \bar{\mathcal{T}}_{\lambda+\mu-\alpha} .$$

If $\lambda + \mu - \alpha \neq 0, 1$ or $\frac{1}{2}$ then

$$[[jr, a], b] \in \bar{\mathcal{T}},$$

which is impssible in view of (28) and (30) .

Using the symmetry of abc , we obtain now

(33) $$\lambda + \mu, \mu + \nu, \nu + \lambda = \alpha, 1 + \alpha \text{ or } \frac{1}{2} + \alpha .$$

Lemma 15 shows that if $\varrho + b > 0$ or $\varrho = b = 0$, then

$$[[\bar{\mathcal{G}}^{(-1)}, \bar{\mathcal{G}}^{(1)}], \bar{\mathcal{G}}_b^{(1)}] = 0 .$$

Since $[\bar{jr}, a] \in \bar{\mathcal{G}}^{(1)}_{\lambda-\alpha}$, $[\bar{jr}, c] \in \mathcal{G}^{(1)}_{\nu-\alpha}$, condition (31) can be satisfied only in the case where $\lambda + \nu - 2\alpha \leq 0$, $\lambda - \alpha$, $\nu - \alpha$ are not simultaneously zero .

Using the symmetry of abc we obtain

(34) $$\lambda + \mu, \mu + \nu, \nu + \lambda \leq 2\alpha ,$$

where at most one of the three numbers

(35) $$\lambda, \mu, \nu \quad \text{is equal to } \alpha .$$

It is easy to see that, just as in the case $\alpha = 0$, also in the case

$\alpha = \frac{1}{2}$ the conditions (32) - (35) cannot be simultaneously satisfied.

6. The statement of lemma 17 can be rephrased as follows

(36) $$[[\mathfrak{g}^{(1)}, \mathfrak{g}], \mathfrak{g}] \subset \mathfrak{g}^{(0)} .$$

From the commutativity of $\mathfrak{g}^{(1)}$ and the Jacobi identity it follows that

(37) $$[\mathfrak{g}^{(2)}, [\mathfrak{g}^{(1)}, \mathfrak{g}]] = 0 .$$

Condition (36) means that

$$[jr, [[\mathfrak{g}^{(1)}, \mathfrak{g}], \mathfrak{g}]] \subset \mathfrak{g}^{(2)} .$$

By (37) this is equivalent to

(38) $$[[\mathfrak{g}^{(1)}, \mathfrak{g}], \mathfrak{g}^{(1)}] \subset \mathfrak{g}^{(2)} .$$

From (37) and from the closure of the form ϱ it also follows that

$$\varrho([[\mathfrak{g}^{(1)}, \mathfrak{g}], \mathfrak{g}^{(1)}], \mathfrak{g}^{(2)}) = 0 .$$

Comparing this with (38) we finally obtain the relation

(39) $$[[\mathfrak{g}^{(1)}, \mathfrak{g}], \mathfrak{g}^{(1)}] = 0 .$$

Lemma 18 : The centralizer $\mathcal{Z}(\mathfrak{g}^{(1)})$ of the sub-algebra $\mathfrak{g}^{(1)}$ in $\mathfrak{g}$ is a Kähler ideal.

Proof. Note that

(40) $$\mathcal{Z}(\mathfrak{g}^{(1)}) \subset \mathfrak{g}^{(0)}$$

since $jr \in \mathfrak{g}^{(1)}$ and $\mathcal{Z}(jr) \subset \mathfrak{g}^{(0)}$.

Equality (39) means that

(41) $$[\mathfrak{g}^{(1)}, \mathfrak{g}] \subset \mathcal{Z}(\mathfrak{g}^{(1)}) .$$

From the Jacobi identity

$$[jr, [\mathcal{Z}(\mathfrak{g}^{(1)}), \mathfrak{g}]] \subset [\mathcal{Z}(\mathfrak{g}^{(1)}), \mathfrak{g}^{(1)}] = 0$$

(here we use the fact that $jr \in \mathfrak{g}^{(1)}$ and therefore $[jr, \mathcal{Z}(\mathfrak{g}^{(1)})] = 0$).

Consequently $[\mathcal{Z}(\mathfrak{g}^{(1)}), \mathfrak{g}] \subset \mathfrak{g}^{(o)}$, and

(42) $$[[\mathcal{Z}(\mathfrak{g}^{(1)}), \mathfrak{g}], \mathfrak{g}^{(1)}] \subset \mathfrak{g}^{(1)} .$$

Furthermore from (41) it follows that

$$[[\mathcal{Z}(\mathfrak{g}^{(1)}), \mathfrak{g}], \mathfrak{g}^{(1)}] \subset [\mathcal{Z}(\mathfrak{g}^{(1)}), \mathcal{Z}(\mathfrak{g}^{(1)})$$

and

(43) $$\rho([[\mathcal{Z}(\mathfrak{g}^{(1)}), \mathfrak{g}], \mathfrak{g}^{(1)}], \mathfrak{g}^{(1)}) \subset$$

$$\subset \rho([\mathcal{Z}(\mathfrak{g}^{(1)}), \mathcal{Z}(\mathfrak{g}^{(1)})], \mathfrak{g}^{(1)}) = 0 .$$

Comparing (42) with (43) we find that

$$[[\mathcal{Z}(\mathfrak{g}^{(1)}), \mathfrak{g}], \mathfrak{g}^{(1)}] = 0, \text{ i.e } [\mathcal{Z}(\mathfrak{g}^{(1)}), \mathfrak{g}] \subset \mathcal{Z}(\mathfrak{g}^{(1)}) .$$

Therefore $\mathcal{Z}(\mathfrak{g}^{(1)})$ is an ideal in $\mathfrak{g}$.

We show now that the ideal $\mathcal{Z}(\mathfrak{g}^{(1)})$ is invariant under j. Let $z \in \mathcal{Z}(\mathfrak{g}^{(1)})$, Then $jz \in \mathcal{Z}(\mathfrak{g}^{(o)})$. From the integrability conditions it follows that the operator $A = \mathrm{ad}_{\mathfrak{g}^{(1)}} jz$ commutes with j, and from the closure of ρ and the commutativity of $\mathfrak{g}^{(1)}$ it follows that it is skew-symmetric with respect to ρ.

Therefore the operator A is skew-symmetric with respect to the canonical scalar product and since it has real eigen-values we have $A = 0$. This means that $jz \in \mathcal{Z}(\mathfrak{g}^{(1)})$. The lemma is proved.

7. Now we can prove proposition 3.

We denote by N the center of the ideal $\mathcal{Z}(\mathfrak{g}^{(1)})$. This will be a commutative ideal in $\mathfrak{g}$ and by lemma 7 applied to the Kähler algebra

$\mathcal{Z}(\mathfrak{g}^{(1)})$ and ideal N is j-invariant. Since $N \supset \mathfrak{g}^{(1)}$ we have $N \neq 0$. So proposition is proved under the induction hypothesis of the fundamental theorem .

From propositions 2 and 3 , which we have proved in §§ 2 and 3 respectively, proposition 1 follows.

§ 4. Proof of the fundamental theorem.

Let $\mathfrak{g}$ be a normal Kähler algebra and assume that the statement of the theorem is true for all normal Kähler algebras of dimension lower than that of $\mathfrak{g}$. Then we can apply propositions 2 and 3 to $\mathfrak{g}$.

Let N be a Kähler ideal in $\mathfrak{g}$ satisfying the conditions of one of these propositions and let $\mathfrak{g}'$ be the orthogonal complement of N. Then $\mathfrak{g}'$ is a Kähler sub-algebra of $\mathfrak{g}$ (cf. § 6, part 2) . By the induction hypothesis $\mathfrak{g}'$ can be decomposed into a semi-direct sum

(47) $$\mathfrak{g}' = \mathcal{J}' + \mathcal{H}' ,$$

where $\mathcal{J}'$ is a commutative Kähler ideal and $\mathcal{H}'$ is a Kähler sub-algebra having a principal idem potent s' .

Applying the lemma on symplectic representations (cf. § 6 , part 2) to the Kähler algebra $N + \mathcal{J}'$ we see that

(48) $$[N, \mathcal{J}'] = 0 .$$

We consider separately the two cases corresponding to the two possible types of the ideal N .

a) N is an elementary Kähler algebra . We set

$$\mathcal{J} = \mathcal{J}' , \quad \mathcal{H} = N + \mathcal{H}' , \quad s = r + s' ,$$

where $r \in N$ is an element fulfilling the conditions of proposition 2. We show that the decomposition $\mathcal{G} = \mathcal{J} + \mathcal{H}$ is of the kind needed for the theorem and s is a principal idem potent for the algebra $\mathcal{H}$.

From (48) it follows that $\mathcal{J}$ is an ideal in $\mathcal{G}$. Since $\mathcal{G}'$ commutes with jr and r we have

$$[js, s] = [jr, r] + [js', s'] = r + s' = s .$$

By the same considerations

$$\text{ad } js = \begin{cases} \text{ad } js' & \text{on } \mathcal{G}' \\ \text{ad } jr & \text{on } (jr) + (r) . \end{cases}$$

It follows now that the operator ad js is semi-simple on the sub-space

$$\mathcal{W} = (jr) + (r) + \mathcal{H}' \subset \mathcal{H}$$

and has eigen-values $0, 1, \frac{1}{2}$ there, and $j\mathcal{W}_{\lambda} = \mathcal{W}_{1-\lambda}$. It remains to check how js acts on the sub-space $\mathcal{U}$ (cf. proposition 2).

Consideration of the eigen-values of ad jr on $\mathcal{G}$ shows that $[\mathcal{G}', \mathcal{U}] \subset \mathcal{U}$.

The representation $\text{ad}_{\mathcal{U}}$ of the two-dimensional Kähler algebra generated by s' and js' will be sumplectic.

By the lemma about the symplectic representations of such algebras (§ 6, part 2)

$$\mathcal{U} = \mathcal{U}'_{-\frac{1}{2}} + \mathcal{U}' + \mathcal{U}'_{0} ,$$

where

$$\text{ad } js' = \lambda \quad \text{on } \mathcal{U}'_{\lambda} ,$$

$$j\,\mathcal{U}'_{\lambda} \subset \mathcal{U}'_{-\lambda} .$$

Since $\mathrm{ad}_{\mathcal{U}}\, jr = \frac{1}{2}$ setting $\mathcal{U}_\lambda = \mathcal{U}'_{\lambda - \frac{1}{2}}$ we obtain

$$\mathcal{U} = \mathcal{U}_0 + \mathcal{U}_1 + \mathcal{U}_{\frac{1}{2}} ,$$

$$\mathrm{ad}\; js = \lambda \quad \text{on } \mathcal{U}_\lambda ,$$

$$j\, \mathcal{U}_\lambda = \mathcal{U}_{1-\lambda} .$$

This finishes the proof in the case a) .

b) N is a commutative Kähler algebra .

We set

$$\mathcal{J} = N + \mathcal{J}' , \mathcal{H} = \mathcal{H}' .$$

From (48) it follows that the ideal $\mathcal{J}$ is commutative so that the decomposition $\mathcal{G} = \mathcal{J} + \mathcal{H}$ satisfies the requirements of the theorem .

Some problem .

In § 5 of the introduction to our lectures the "main conjecture" about the structure of the homogeneous Kählerian manifolds (h.K.m.) was formulated. We give some corollaries of it here. It is possible that some of them are valid for a wider class of manifolds. It should be very interesting to find direct proofs for these statements.

1. Each h.K.m. whose points are separated by bounded holompr-phic functions is a bounded domain.

2. If the points of a h.K.m. are separated by holomorphic functions but all bounded holomorphic functions on it are constant then it is locally flat.

4. Each simply connected h.K.m. is holomorphically convex.

5. If a h.K.m. is a Stein manifold then it may be holomorphically fibered in locally flat h.K.m. and the base is a bounded domain.

For arbitrary holomorphically convex complex manifolds the Remmert fibration is known, its fibres being defined as maximal sets on which all holomorphic functions are constant. The base of this fibration is a Stein manifold and the fibres are compact. This result could be applied for the proof of the main conjecture if the propositions 4 and 5 were proved. The similar construction for bounded holomorphic functions, about which we do not know any general results, gives immediately the fibration mentioned in the main conjecture.

6. Each h.K.m. with negative Ricci curvature is a bounded domain.

7. Each h.K.m. with zero Ricci curvature is locally flat.

Hano and Kobayashi considered one canonical fibering of an arbitrary homogeneous complex manifold with invariant measure. It seems to be not difficult to prove that for h.K.m. the fibres of this fibering have zero Ricci curvature and the base is a h.K.m. with non-degenerate Ricci curvature. It is possible that a further investigation of the base and a direct proof of the statement 7 will lead to a differential-geometrical proof of the main conjecture.

8. Each compact group of automorphisms of a h.K.m. has an orbit which is a complex submanifold. In particular, each one-parametric compact group of automorphisms has a fixed point.

REFERENCES

1. BERGMANN S. , "Über die Kernfunctionen eines Bereiches und ihr Verhalten am Rande", J. reine und angew. Math., 1933, 169 , 1934, 172, 89-128 .

2. BOREL A., "Kählerian coset spaces of semisimple Lie groups" , Proc. Nat. Acad. Sci. U.S.A. , 1954 , 40 , 12, 1147-1151.

3. BOREL A., REMMERT R. , "Über kompakte homogene Kählersche Mannigfaltigkeiten", Math. Ann. , 1962 , 145, 5, 429-439 .

4. CARTAN E. , "Sur les domaines bornés homogènes de l'espace de n variables complexes" , Abh. Math. Sem. Hamb. Univ., 1935, II, 116-162.

5. FUKS B.A. , " pecial chapters in the theory of analytic functions of several complex variables", Moscow, 1963 (Russian), (English Translation published by the American Mathematical Society in 1965) .

6. GRAUERT H., "Analytischen Faserungen über holomorf-vollständigen Räumen", Math. Ann. , 135, 3, 263-273.

7. HANO J. , "On Kahlerian homogeneous spaces of unimodular Lie groups", Amer. J. Math., 1957 , 79, 4, 885-900.

8. KOBAYASHI S. , NOMIZU K. , "On automorphisms of a Kählerian structure" , Nagoya Math. J., 1957 , II, 115-124 .

9. KOSZUL J.L. , " Sur la forme hermitienne canonique des espaces homogenes complexes", Canad. Journ. Math. 7,4 , 1955, 562-576.

10. KOSZUL J.L. , "Ouverts convexes des espaces affines", Math. Zeitschr. 79, 1962, 254-259.

11. LICHNEROWICZ A. , "Espaces homogenes kähleriennes", Colloque de geometrie differentielle", Strasbourg, 1953, 171-184.

12. LICHNEROWICZ A. , "Sur les groupes d'automorfismes de certaines variétés Kählériennes" , C.r. Acad. Sci. Paris, 1954, 239, 21 , 1344-1346.

13. LICHNEROWICZ A. , "Theorie globale des connexions et des groups d'holonomie", Rome, 1955.

14. MATSUSHIMA Y. Sur les espaces homogènes Kähleriens d'un groupe de Lie reductif" , Nagoya Math. J. , 1957, 53-60.

15. POINCARÉ H. , "Les fonctions analytiques de deux variables et la representation conforme", Rend. Circolo Mat. di Palermo, 1907, 23, 185-220.

16. PJATECCKIĬ, ŠAPIRO I.I. , "On a problem proposed by E. Cartan", Dokl. Akad. Nauk. SSSR. 113 (1957) , 980-983 (Russian) .

17. PJATECCKIĬ, ŠAPIRO I.., "Geometry of classical domains and theory of automorphic functions", Moscow 1961 (Russian) (an enlarged english edition is being prepared for print) .

18. PJATECCKIĬ, ŠAPIRO I.I., "Classification of bounded homogeneous domains in n - dimensional complex space", Dokl. Akad. Nauk. SSSR, 141, 2 (1961) , 316-319 (Russian) = Soviet Math. Dokl. 2(1961) , 1460-1463.

19. PJATECCKIĬ, ŠAPIRO I.I. , "On bounded homogeneous domains in an n - dimensional complex space", Izv. Akad. Nauk SSSR, Ser. Mat. 26 (1962) , 107 - 124 (Russian) .

20. PJATECCKIĬ, ŠAPIRO I.I. , The structure of j algebras" , ibd. 26 (1962) , 453-484.

21. PJATECCKIĬ, ŠAPIRO I.I , "The geometry and classification of bounded homogeneous domains", Uspehi Mat. Nauk. 20 (1965), no. 2, 3 - 51 (Russian) = Russian Math. Surveys, 20 (1965) , no. 2, 1 -48.

22. TITS J. , "Espaces homogènes complexes compacts" , Comm. Math. Helv., 1962, 37, 2, 111 - 120 .

23. VINBERG E.B., "The Morozov - Borel theorem for real Lie groups" , Dokl. Akad. Nauk. SSR 141, 2 (1961) , 270-273 (Russian - Soviet Math. Dokl. 2 (1961) , 1416 - 1419 .

24. VINBERG E.B., "The theory of convex homogeneous cones" , Trady Moskov. Math. Obshch. 12 (1963) , 303-358 (Russian) = Trans. Moscow Math. Soc. 13 (1964) , 340-403.

25. VINBERG E.B. , GINDIKIN S.G., PJATECCKII, SAPIRO I.I., "On the classification and canonical realization of complex homogeneous bounded domains", ibd., 359-388 (Russian) = ibd. , 404-437.

26. WANG H.C. , "Closed manifolds with homogeneous complex structure" Amer. J. Math. , 1954, 76, I, 1-32.

27. WEIL A. , "Introduction à l'étude des variétés kählériennes" , Paris, 1958.

CENTRO INTERNAZIONALE MATEMATICO ESTIVO

(C.I.M.E.)

Stephen J. GREENFIELD

EXTENDIBILITY PROPERTIES OF REAL SUBMANIFOLDS OF $\mathbb{C}^n$

Corso tenuto ad Urbino dal 5 al 13 luglio 1967

EXTENDIBILITY PROPERTIES OF REAL SUBMANIFOLDS OF $\mathbb{C}^n$

by

Stephen J. Greenfield
(Massachusetts Institute of Technology Cambridge, Mass)

A History

In 1906 F. Hartogs [4] discovered that a function analytic in a neighborhood of the bicyclinder in $\mathbb{C}^2$ could always be extended to an analytic function defined in a neighborhood of all the bicyclinder. Not much later (1910) E. E. Levi [6] found a local analogue of Hartogs' result : let $h: \mathbb{C}^2 \longrightarrow R$ be differentiable and suppose that $M = h^{-1}(0)$ is a submanifold of $\mathbb{C}^2$. If

$$(1) \qquad \det \begin{pmatrix} h_{z_1\bar{z}_1} & h_{z_2\bar{z}_1} & h_{\bar{z}_1} \\ h_{z_1\bar{z}_2} & h_{z_2\bar{z}_2} & h_{\bar{z}_2} \\ h_{z_1} & h_{z_2} & 0 \end{pmatrix} \neq 0$$

at $p \in M$, then functions analytic in any neighborhood of M extend to be analytic in a fixed open set on one side of M. The boundary of the fixed open set includes a neighborhood of p in M.

In the early 1940's, Bochner and Martinelli [2] [9] showed that if M is a compact differentiable hypersurface bounding an open set $\mathcal{U}$ of $\mathbb{C}^n$, then any function analytic in a neithborhood of M has an extension analytic in $\mathcal{U}$. (Their result was actually better, for they required only that the function be defined on M and satisfy certain appropriate partial differential equations.)

In 1960, H. Lewy [8] published an example of a four-dimensional manifold in $\mathbb{C}^3$ having the property that functions analytic in a neighborhood of it extend to be analytic in a fixed open set. (As in the preceding, Lewy just needed the function defined on the manifold and satisfying a differential equation). Lewy's investigation of extension had an important byproduct [7] . The first example of what he calls 'atypical' partial differential equations was discovered; these equations have no solution in any open set.

E. Bishop (1965) [1] gave a general method of considering four-dimensional manifolds in $\mathbb{C}^3$ and discovering when they had the 'extension property' to open subsets of $\mathbb{C}^3$. His work was generalized by B. Weinstock [13] to submanifolds of real codimension 2 in $\mathbb{C}^n$ (1966) . Other recent work on extendibility has been done by H. Rossi [12] and R.O. Wells [14-16] .

B Some definitions from several complex variables.

Let K be a subset of $\mathbb{C}^n$. H(K) is the collection of all functions f defined and analytic in a neighborhood of K . We say that K is extendible to a connected subset K' of C^n (containing K) if res : H(K') $\longrightarrow$ H(K) (the natural restriction map) is onto.

If K is a subset of $\mathcal{U}$, open in $\mathbb{C}^n$, then the $H(\mathcal{U})$ hull of K, $\hat{K}_{\mathcal{U}}$, is $\{p \in \mathcal{U} \,\big|\, |f(p)| \leqslant \sup_K |f|,\ f \in H(\mathcal{U})\}$.
A subset L of $\mathcal{U}$ is $H(\mathcal{U})$ - convex when : K compact subset of L $\Longrightarrow \hat{K}_{\mathcal{U}}$ compact subset of L .

A open set $\mathcal{U}$ of $\mathbb{C}^n$ is a domain of holomorphy only when it is $H(\mathcal{U})$ convex.

C Statement of the problem, and an investigation of the hypersurface case .

A problem of several complex variables, of importance also to function algebras and partial differential equations, is :
given two subsets of $\mathbb{C}^n$, K and L (with $K \subset L$), when is K extendible to L? Here we investigate this problem when K and L are submanifolds

of $\mathbb{C}^n$. How big (dimension) can L be in terms of K ? For K of high codimension, can L be open ? In what sense is K part of the boundary of L ?

Bishop has written: "It is thought that a manifold $M^{n+1} \subset \mathbb{C}^n$ has, in general, the property that holomorphic functions in a neighborhood of M extend to be holomorphic in some fixed open set. We will prove this, and discover what "in general" means.

First we examine the case of a hypersurface M of $\mathbb{C}^n$ carefully. Since $\mathbb{C}^n$ has a complex structure, the complexified tangent bundle splits as the direct sum of equal-dimensional subbundles :

$T(\mathbb{C}^n) \otimes \mathbb{C} = H(\mathbb{C}^n) + A(\mathbb{C}^n)$.

At $p \in \mathbb{C}^n$, $H(\mathbb{C}^n)$ (the holomorphic tangent bundle) is generated by $\left(\frac{\partial}{\partial z_j}\right)_p$, $1 \leqslant j \leqslant n$. The antiholomorphic tangent bundle, $A(\mathbb{C}^n)$ is generated by $\left(\frac{\partial}{\partial \bar{z}_j}\right)_p$

We know :

$$H(\mathbb{C}^n) \cap A(\mathbb{C}^n) = 0 \ , \quad [\Gamma H(\mathbb{C}^n) \ , \ \Gamma H(\mathbb{C}^n)] \subset \Gamma H(\mathbb{C}^n) \ .$$

(If V is a vector bundle, ΓV will denote the C^∞ sections of V.)

Consider now $T(M) \otimes \mathbb{C} \subset T(\mathbb{C}^n)\big|_M \otimes \mathbb{C}$. Define the holomorphic (resp. antiholomorphic) tangent bundles of M by $H(M) = H(\mathbb{C}^n) \cap T(M) \otimes \mathbb{C}$ (resp. $A(M) = A(\mathbb{C}^n) \cap T(M) \otimes \mathbb{C}$) . Then we have these properties :

(2) $$H(M) \cap A(M) = 0 \ , \quad [\Gamma H(M) \ , \ \Gamma H(M)] \subset \Gamma H(M) \ .$$

Since the Lie bracket is $\overline{}$- invariant in $T(M) \otimes \mathbb{C}$, we know also that $[\Gamma A(M) \ , \ \Gamma A(M)] \subset \Gamma A(M)$. But $H(M) + A(M)$ is not all of $T(M) \otimes \mathbb{C}$. In fact, $H(M) + A(M)$ has codimension 1 in $T(M) \otimes \mathbb{C}$.

It is not too difficult to show that condition (1) of E. E. Levi is equivalent to

(3) $$\left[\Gamma H(M),\ \Gamma A(M)\right] \not\subset \Gamma(H(M) + A(M))\ .$$

In fact, M is extendible (to an open set of $\mathbb{C}^n$) only when (3) is true . This gives us ample encouragement to initiate a theory for higher codimension.

D C - R manifolds

Definition : A C - R manifold is a pair (M, H(M)) where M is a real differentiable manifold of dimension n + k (n $\geqslant$ k) and H(M) is a k-dimensional complex subbundle of T(M) $\otimes\mathbb{C}$ so that (if $A(M) = \overline{H(M)}$)

(4) $$H(M) \cap A(M) = 0, \quad \left[\Gamma H(M)\ ,\ \Gamma H(M)\right] \subset \Gamma H(M)$$

These conditions are adapted to imitate (2) , and of course we shall call H(M)) the holomorphic (resp. antiholomorphic) tangent bundle of M.

There are numerous equivalences for the conditions of (4) .

In particular we can get :

Proposition: If M is a real differentiable manifold of dimension n+k , then the first conditions is the same as :

There is a (2 k)-dimensional subbundle R of T(M) so that R is a complex vector bundle (that is, there is a real bundle map J : R $\rightarrow$ R so that $J^2 = -Id_R$) .

The second condtion is equivalent to :

$[a, b] + J\,[Ja, b] + J\,[a, Jb] - [Ja, Jb] = 0$ for any a, b$\in \Gamma$R . The vanishing of the Nijenhuis tensor for C-R structure .)

Proof: Put R = re (H(M) + A(M)) , and J can be obtained easily . The second equivalence is a computation.

The C-R codimension of M is the fiber $\dim_{\mathbb{R}}(T(M)/R)$. (We shall assume M connected, so this number is well-defined) .

Examples : a) If C-R codim M = 0 , then the first condition of (4) says that M is almost-complex. The second, by the Newlander-Niremberg theorem [10] , implies that M is a complex manifold.

b) Suppose $P = \sum_1^n a_j \frac{\partial}{\partial x_j}$ is a partial differential operator on $\mathbb{C}^n$ with C^∞ complex-valued coefficients, and that P and $\bar{P}$ are always linearly independent (over $\mathbb{C}$) . Then P determines a C-R structure on $\mathbb{R}^n$, with C-R codim = n - 2 .

c) (a non-example) S^4 has no C-R structure, in any C-R codim. The proof uses equipment of algebraic topology.

d) Contact manifolds are C-R manifolds with C-R codim = 1 There are many other examples.

In the case of complex manifolds, the $\bar{\partial}$ complex, with its accompanying Dolbeault cohomology groups, is well-known.

We won't go into details, but merely mention the following : let M be a C-R manifold. If $\mathcal{D}^{p,q} = \Gamma(\Lambda^p H(M)^* \otimes \Lambda^g A(M)^*)$, then :

Theorem: $\bar{\partial} : \mathcal{D}^{o,g} \to \mathcal{D}^{o,g+1}$ is well-defined.

Theorem : $\bar{\partial}^2 = 0$, and this statement is equivalent to the second condition of (4) .

The cohomology groups defined implicitly above have been studied by J. J. Kohn in the case that M is compact and C-R codim m = 1 . He has obtained some very interesting results [5] .

Guided by (3) we want to consider ' cross-terms' obtained by bracketing $\Gamma H(M)$ and $\Gamma A(M)$. So we define the Levi algebra of M , $\mathcal{L}(M)$, to be the Lie subalgebra of $\Gamma T(M) \otimes \mathbb{C}$ generated by $\Gamma H(M)$ and $\Gamma A(M)$. We will assume that $\mathcal{L}(M)$ has constant dimension at each point of M. It is clearly a $\bar{\ }$-invariant subalgebra of $\Gamma T(M)$, so there is a complex vector subbundle V of $T(M) \otimes \mathbb{C}$ with $\Gamma V = \mathcal{L}(M)$. We define the excess dimension of $\mathcal{L}(M)$, ex dim $\mathcal{L}(M)$, to be the fiber $\dim_{\mathbb{C}} (V/(H(M) + A(M)))$.

If $M = N \times T$, where N is a complex manifold, and T is a real manifold, and if $H(M) = H(N) \times 0$, then ex dim $\mathcal{L}(M) = 0$. The converse is true locally by the (non-trivial) complex Frobenius theorem of L. Niremberg [11] :

Theorem : Suppose $(M, H(M))$ is a C-R manifold with ex dim $\mathcal{L}(M) = 0$. If $p \in M$, there is an open neighborhood of p in M which is C-R isomorphic to an open neighborhood of 0 in $\mathbb{R}^{h-k} \times \mathbb{C}^k$ (with the natural C-R structure on the latter) .

(Note: Since we used the phrase 'C-R isomorphism' above the concept of a C-R map should be mentioned. If $(M, H(M))$ and $(N, H(N))$ are C-R manifolds, a C-R map $f: M \longrightarrow N$ is a differentiable map so that $df \otimes 1(H(M)) \subset H(N)$.
Equivalently, df commutes with 'J' or Λf (the natural map of exterior algebras) commutes with '$\bar{\partial}$' . When M and N are manifolds, the notion of C-R map coincides with that of complex analytic map-f is required to satisfy the appropriate Cauchy-Riemann equations.)

E Embedded C - R manifolds, and generic manifolds

Suppose M is a submanifold of $\mathbb{C}^n$. Then $H(M)$ is not necessarily a vector bundle. If $H(M)$ is a vector bundle, then $(M, H(M))$ is a C-R submanifold of $(\mathbb{C}^n, H(\mathbb{C}^n))$. But how can we guarantee that H(M) is a vector bundle ?

$T(M)_p$ (by affine translation) is a real linear subspace of $\mathbb{C}^n$.

$H(M)_p$ is canonically isomorphic to the largest complex vector subspace of $\mathbb{C}^n$ contained in $T(M)_p$. With a short digression to linear algebra, we can obtain a large supply of 'good' M's .

Linear algebra: Let W be a complex vector space of complex dimension n , and V be a real vector subspace of W of real dimension k . Let m(V) denote the maximal complex subspace of W in V. Then :

$$\max(0, k-n) \leqslant \dim_{\mathbb{C}} (m(V)) \leqslant \frac{k}{2} .$$

Let $G_{\mathbb{R}}^{p}(W)$ (resp. $G_{\mathbb{C}}^{p}(W)$) be the p-dimensional real (resp. complex) Grassmanians of W . Put $G_{\mathbb{C}}(W) = G_{\mathbb{C}}^{0}(W) + G_{\mathbb{C}}^{1}(W) + \ldots + G_{\mathbb{C}}^{n}(W)$.
If $V \in G_{\mathbb{R}}^{k}(W)$, then $m(V) \in G_{\mathbb{C}}(W)$.

Theorem : If p = max (0, k-n) , then $m : G_{\mathbb{R}}^{k}(W) \longrightarrow G_{\mathbb{C}}(W)$ has the following properties :

a) $m^{-1}(G_{\mathbb{C}}^{p}(W))$ is a dense open subset of $G_{\mathbb{R}}^{k}(W)$.

b) $m \big|_{m^{-1}(G_{\mathbb{C}}^{p}(W))}$ is a C^{∞} map .

Proof : Take coordinates, and use an argument based on ranks of matrices.

Elements of $m^{-1}(G_{\mathbb{C}}^{p}(W))$ are called generic subspaces of dimension k and other elements of $G_{\mathbb{R}}^{k}(W)$ are called exceptional subspaces.

If $T(M)_p$ is a generic subspace of $\mathbb{C}^n$, we call p a generic point of M . If p is generic, an open neighborhood N of p in M consists of generic points, and (N, H(N)) is a C-R manifold, called a generic C-R submanifold of $\mathbb{C}^n$. If $H(N) \neq 0$, observe that C-R codim $N = \text{codim}_{\mathbb{R}}$ in $\mathbb{C}^n$.

We give a general example of non-trivial generic submanifolds of $\mathbb{C}^n$ with $\dim_{\mathbb{R}} = n+k$. Let $r_1, \ldots, r_{n-k}$ be C^∞ real-valued functions on $\mathbb{C}^n$. Suppose $p \in \bigcap_j r_j^{-1}(0)$. If $dr_1(p) \wedge \ldots \wedge dr_{n-k}(p) \neq$, there is an open neighborhood $\mathcal{U}$ of p in $\mathbb{C}^n$ so that $\mathcal{U} \bigcap_j r_j^{-1}(0) = M$ is a C^∞ submanifold of $\mathbb{C}^n$ of codimension n-k . If, in addition, $\bar{\partial} r_1(p) \wedge \ldots \wedge \bar{\partial} r_{n-k}(p) \neq 0$ (the r_j are holomorphically transverse at p) then $\mathcal{U}$ can be chosen so that M is a generic C-R submanifold of $\mathbb{C}^n$. ($r_1 = x_1$, $r_2 = y_1$ in ^{n} gives on example of a non-generic submanifold, for $\bar{\partial} r_1 \wedge \bar{\partial} r_2 = 0$).

Theorem: If M is a generic C-R submanifold of $\mathbb{C}^n$ given in the above manner by transverse, holomorphically transverse functions, then $f : M \longrightarrow \mathbb{C}$ is a C-R map only when

$$\bar{\partial} f \wedge \bar{\partial} r_1 \wedge \ldots \wedge \bar{\partial} r_{n-k} = 0 .$$

f is said to be "relatively holomorphic" and the partial differential equations above are usually called the "induced" or "tangential" Cauchy-Riemann equations. Note that the restriction to M of any function holomorphic in a neighborhood of M satisfies these differential equations, and so is a C-R map .

All real hypersurfaces of $\mathbb{C}^n$ are generic. A C-R submanifold of codimension 2 is either or a complex analytic hypersurface.

The extendibility questions can be answered most completely for generic submanifolds of $\mathbb{C}^n$ with non-trivial holomorphic tangent bundle . Therefore we restrict attention to this case .

When ex dim $\mathcal{L}(M) = 0$, we have more information than was given by the theorem in D [3] .

<u>Theorem</u>: Let M be a generic C-R submanifold of $\mathbb{C}^n$ with $H(M) \neq 0$. The following are equivalent for $p \in M$:.

a) There is a fundamental system of neighborhoods $\mathcal{U}$ of $p \in M$ so that ex dim $\mathcal{L}(\mathcal{U}) = 0$

b) There is a fundamental system of neighborhoods $\mathcal{U}$ of $p \in M$ so that $\mathcal{U}$ is C-R isomorphic to $\mathbb{R}^s \times \mathbb{C}^q$, with s = C-R codim M .

c) There is a fundamental system of neighborhoods $\mathcal{U}$ of $p \in M$ so that $\mathcal{U} = \bigcap_{j \in \mathbb{Z}} S_j$, S_j domains of holomorphy .

d) There is a fundamental system of neighborhoods $\mathcal{U}$ of $p \in M$ so that $\mathcal{U}$ is not extendible .

e) For sufficiently small open balls B of $\mathbb{C}^n$ with center p , $B \cap M$ is H(B) convex .

When ex dim $\mathcal{L}(M) = 0$, M is analogous to a domain of holomorphy . When ex dim $\mathcal{L}(M) > 0$, we will have interesting extendibility .

F Bishop discs

<u>Definition</u>: A family of analytic discs in $\mathbb{C}^n$ parameterized by a subset $\mathcal{U}$ of $\mathbb{R}^n$ is a map $F : \mathcal{U} \times D \to \mathbb{C}^n$ (where D is the subset $\{|z| < 1\}$ of $\mathbb{C}$) so that F(t, -) is analytic as a map of D into $\mathbb{C}^n$ for any t .

There is a classical Kontinuitässatz formulated by R.O. Wells [14] :

Theorem : Let $\mathcal{U}$ be a simply connected domain in $\mathbb{R}^d$.
If $F : \mathcal{U} \times \bar{D} \longrightarrow \mathbb{C}^n$ is a continuous family of analytic discs parameterized by $\mathcal{U}$ so that, for some $u_0 \in \mathcal{U}$, $F(\{u_0\} \times D)$ is a point (a degenerate disc), then $F(\mathcal{U} \times \partial D)$ is extendible to $F(\mathcal{U} \times \bar{D})$.

Bishop used the theorem above to investigate extendibility problems of submanifolds by creating families of analytic discs containing degenerate discs so that :

a) The boundaries of the discs are contained in the given submanifold.

b) The interior of the family (that is, $F(\mathcal{U} \times D)$) "fills up" an open set in a manifold of higher dimension.

a) and b) will be called, respectively, the problems of existence and non-triviality for a family of analytic discs. Bishop investigated these problems for an M^4 in $\mathbb{C}^3$, and showed that with suitable assumptions (corresponding to generic with ex dim $\mathcal{L}(M) = 2$) the existence problem could be solved with a continuous family of discs, and the discs were non-trivial : their interiors contained an open subsubset of $\mathbb{C}^3$.

The existence problem is not simple. It involves the solution of partial differential equations with strict boundary conditions. For generic submanifolds in $\mathbb{C}^n$, Weinstock, generalizing Bishop's work, showed the existence of a large collection of differentiable families of analytic discs. He used this to show extendibility from submanifolds of codimension 2 in $\mathbb{C}^n$ to open sets in $\mathbb{C}^n$ (under the same conditions as Bishop : generic and ex dim $\mathcal{L}(M) = 2$).

For higher codimension, using what is in some respects an actual simplification of the Bishop-Weinstock proof, we can obtain [3] :

Theorem : Let M be a generic C-R submanifold of $\mathbb{C}^n$ of real dimension $n+k$. Suppose ex dim $\mathcal{L}(M) \geq t > 0$. If $p \in M$, then for all sufficiently small balls B_p in $\mathbb{C}^n$ with center at p, there is a generic C-R submanifold N of B_p of real dimension $n+k+1$, with ex dim $\mathcal{L}(N) \geq t-1$ and $M \cap B_p \subseteq \partial N$. And N will be given as a subset of the regular set of a map $F: \mathbb{R}^{n+k-1} \times \bar{D} \longrightarrow \mathbb{C}^n$ so that

a) F is analytic in the second factor.

b) $F(\mathbb{R}^{n+k-1} \times \partial D) \subset M$.

c) F contains degenerate discs.

Remarks on the proof: Select $u \in \Gamma H(M)$ so that $[u, \bar{u}]_p \notin H(M)_p + A(M)_p$.

Then $i[u, \bar{u}]_p \notin T(M)_p \otimes \mathbb{C}$, and we use the Weinstock theorem, specifying (essentially) that the centers of our discs will lie in the $i[u, \bar{u}]_p$ direction. These discs will be non-trivial. The generic condition of M insures that some subset of the interior of the family is generic.

Finally it is shown without too much trouble that the excess dimension of the Levi algebra of the new generic manifold behaves correctly.

A very geometric example will perhaps make the theorem (and the idea of the proof) clearer. Consider $y_2 = |z_1|^2$ in $\mathbb{C}^2$, where $z_k = x_k + iy_k$. In this simple case a disc whose boundary is on M can be given precisely. Indeed, a full family of discs parameterized by $\mathbb{R}^2$ is :

$$(x_2, y_2, R, \theta) \xrightarrow{F} \begin{cases} z_1 = R\sqrt{y_2}\, e^{i\theta} \\ z_2 = x_2 + i y_2 \end{cases}$$

F is nontrivial ; its Jacobian is $2\, y_2\, R$.

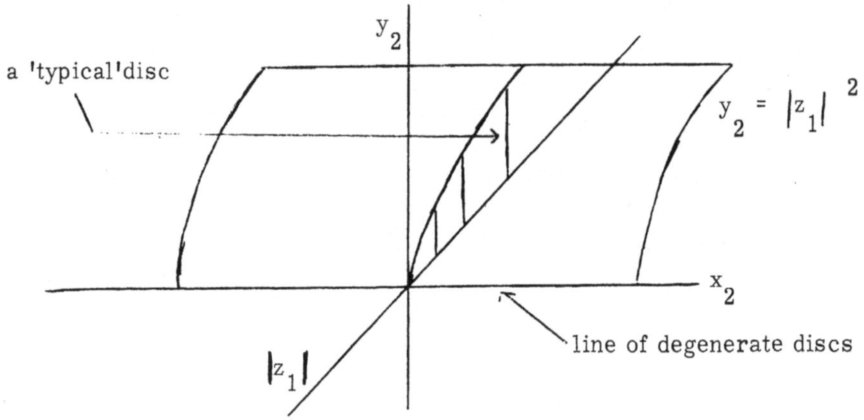

G Iteration and some theorems

If we apply the second theorem in F repeatedly, and then use the Kontinuitätssatz, we obtain :

1 Theorem : If M is a generic C-R submanifold of $\mathbb{C}^n$, and $e = \text{ex dim}\ \mathcal{L}(M) > 0$, then M is extendible to a set containing a generic submanifold N with dim N = dim M + e .

In a certain sense M is a 'boundary' of N (function algebras) .

We can also get a result analogous to the Hartogs-Bochner-Martinelli results for a compact hypersurface :

2 Theorem : Let M be a compact generic C-R submanifold of $\mathbb{C}^n$ with $H(M) \neq 0$. Then M is extendible to a set containing a submanifold N with $\dim N = \dim M + 1$.

Proof: By the previous theorem, if M is not extendible, we must assume ex dim $\mathcal{L}(M) = 0$. Then computation shows that the distribution $\Gamma(\mathrm{re}(H(M) + A(M)))$ is completely integrable and its maximal integral submanifolds possess the structure of non-trivial complex submanifolds of $\mathbb{C}^n$. Then the result desired follows from :

Lemma: No compact subset of $\mathbb{C}^n$ is the union of the images of non-constant analytic maps from connected complex manifolds.

(I know three proofs of this, all depending more or less on the maximum principle. I leave it as an exercise.)

A C-R manifold is called 0-complex if it possesses no imbedded complex submanifolds. 0-complex manifolds are higher codimensional manifolds with properties similar to strictly pseudoconvex hypersurfaces.

A result which includes E.E. Levi's original theorem follows.

3 Theorem: If M is a 0-complex generic C-R submanifold of $\mathbb{C}^n$ with $H(M) \neq 0$, then M is extendible to a set containing a submanifold N, with $\dim N = \dim M{+}1$.

Proof : Easy, but needs some equipment that has not been mentioned here .

Notes: 2 and 3 are 'best possible' in that M may not be extendible to more than one higher dimension.

In 2, M may not be N (there may be cobordism obstruction). But we can always arrange that $\partial N \cap M \neq \emptyset$.

H Reinhardt manifolds

If K is a subset of $\mathbb{C}^n$, let $K^* = \{z \in K \mid$ none of the coordinates of z is $0\} = K \cap \{z \mid z_1 \dots z_n \neq 0\}$. We define a differentiable map of maximal rank $L : \mathbb{C}^{n*} \to R^n$ by
$L((z_1,\dots,z_n)) = (\log |z_1|, \dots, \log |z_n|)$.

If $p \in \mathbb{R}^n$, $L^{-1}(p)$ is a generic submanifold of $\mathbb{C}^n$ of dimension n. (It is an n-torus, $S^1 X \dots X S^1$ (n tines) topologically). Then :

Lemma : If M is a submanifold of $\mathbb{R}^n$ of dimension k, $L^{-1}(M)$ is a generic C-R submanifold of $\mathbb{C}^n$, of C-R codimension n-k.

$L^{-1}(M)$ is called a Reinhardt manifold.

Extendibility problems for Reinhardt manifolds are easily handled because of the followings result. (If $K \subset \mathbb{R}^n$, ch K will denote the convex hull of K).

Theorem : Suppose K is an arcwise connected subset of $\mathbb{R}^n$. Then $L^{-1}(K)$ is extendible to $L^{-1}(\text{ch } K)$. And if $L^{-1}(K)$ is extendible to $\mathcal{U}$, then $\mathcal{U} \subseteq L^{-1}(\text{ch } K)$. ([3]).

Using this theorem, we can construct examples of theorems 1 - 3 of G in any range of codimension and dimension. In particular if M is a curve with convex hull all of $\mathbb{R}^n$, then $L^{-1}(M)$ is an (n+1) - dimensional Reinhardt submanifold which is extendible to all of $\mathbb{C}^{n*}$. (We say: the holomorphic hull is all of $\mathbb{C}^{n*}$.)

If M is any curve in R^n, the coefficients of the

'Frenet formulas' of M can be used to describe the relations between the two generators of $\mathcal{X}(L^{-1}(M))$.

A straight line segment S in $\mathbb{R}^n$ is convex (S = ch S) . So by the theorem in E and the result above, $L^{-1}(S)$ is not extendible.

$L^{-1}(S)$ is therefore a local product of a real and complex manifold. In fact, computation shows that $L^{-1}(S)$ is an (n-1) parameter family of Riemann surfaces (which are either annuli or infinite strips in the plane) .

Then it is not hard to see that a Reinhardt manifold $L^{-1}(M)$ is 0-complex only when M contains no straight line segments.

We can continue this correspondance, and our theorems on holomorphic hulls will become simple theorems on convex hulls. In particular, we get :

From G 3 : if a submanifold M of $\mathbb{R}^n$ contains no straight line segments, ch M contains an open subset of a manifold N with dim N = dim M + 1 .

G 2 becomes : if M is a compact submanifold of $\mathbb{R}^n$, ch M contains an open subset of a manifold N with dim N = dim M + 1 .

I Problems

I will mention only a few of the numerous problems outstanding .

a) Suppose M is a C-R submanifold of $\mathbb{C}^n$. If f: $M \to \mathbb{C}$ is a C-R map, is f the restriction to M of a map holomorphic in a neighborhood of M ?

b) Is theorem G 1 true for non-generic M ?

c) What is the geometric and analytic significance of the Kohn cohomology groups mentioned in D?

d) If $\mathcal{U}$ is an open subset of $\mathbb{C}^n$, there is a Stein manifold $\hat{\mathcal{U}}$ spread over $\mathbb{C}^n$, and $\hat{\mathcal{U}}$ is the 'largest' set of extendibility for $\mathcal{U}$: What can be said (imitating this) for real submanifolds of $\mathbb{C}^n$? (See [3] and [16]).

e) In G 1 , if M is extendible to a set containing a manifold N , is dim N always less than or equal to dim M + e ?

f) Can every C-R manifold be locally C-R imbedded in $\mathbb{C}^n$? (If so, we can imbed it locally as a generic C-R submanifold of some $\mathbb{C}^n$. Global imbedding is too much to require - if N is a compact complex manifold then $N \times T$ (see D) cannot be globally imbedded.)

REFERENCES

1. Bishop, E. , "Differentiable manifolds in complex Euclidean space", Duke Math. J. n (1965) , 1-22

2. Bochner, S., "Analutic and meromorphic continuation by means of Green's formula" , Ann. Math. 39 (1938) , 14-19 .

3. Greenfield, S., Cauchy-Riemann Equations in Several Variables (Bradeis Univ. thesis, 1967).

4. Hartogs, F. , "Einige Folgerungen aus Cauchyschen Intergralformel bei Funktionen mehrer Verändeṙlichen"Silzb Münchener Akad. , 36 (1906) , 223.

5. Kohn, J.J., "Boundaries of complex manifolds", Proceedings of the Conference on Complex Analysis (Springer-Verlag New York Inc. , 1965) .

6. Levi , E.E. , "Studii sui punti singolari essenziali delle funzioni di due o più variabili complesse", Annali di Mat. Pura ed appl., 3(1910) 61-87 .

7. Lewy, H., "On the local character of the solution of an atypical linear differential equation in three variables and a related theorem for regular functions of two complex variables", Ann. Math., 64(1956), 514-522.

8. ------, "On hulls of holomorphy", Comm. Pure Appl. Math. , 13(1960), 587-591.

9. Martinelli, E., "Alcuni teoremi integrali per le funzioni analitiche di più variabili complesse" , Rend. Accad. Italia, 9(1939), 269-300.

10. Newlander, A., and Nirenberg, L. , "Complex analytic coordinates in almost complex manifolds", Ann. Math. , 65(1957), 391-404 .

11. Niremberg, L., "A complex Frobenius theorem", Seminars on Analytic Functions (Institute for Advanced Study-United States Air Force Office of Scientific Research, 1957) .

12. Rossi, H. , report to appear in the Proceedings of the international Congress of Mathematicians (Moscow, 1966) .

13. Weinstock, B. , On Holomorphic Extension from Real Submanifolds of Complex Euclidean Space (M.I.T. thesis, 1966) .

14. Wells, R.O. , "On the local holomorphic hull of a real submanifold in several complex variables", Comm. Pure Appl. Math. , 19(1966, 145- 165.

15. ----------, "Holomorphic approximation on real-analytic submanifolds of a complex manifold", Proc. A.M.S. , 17(1966), 1272-1275.

16. ----------, "Holomorphic hulls and holomorphic convexity of differentiable submanifolds", to appear Trans. A.M.S.

CENTRO INTERNAZIONALE MATEMATICO ESTIVO

(C. I. M. E.)

W. KAUP

HOLOMORPHE ABBILDUNGEN IN HYPERBOLISCHE RÄUME

Corso tenuto ad Urbino dal 5 al 13 luglio 1967

HOLOMORPHE ABBILDUNGEN IN HYPERBOLISCHE RÄUME.

W. Kaup (ERLANGEN)

Es sei $\mathcal{K}$ die Kategorie aller reduzierten komplexen Räume (vergl. [7]) und Hol (X, Y) die Menge aller holomorphen Abbildungen eines komplexen Raumes X in einen komplexen Raum Y. Ist T ein topologischer Raum und d eine stetige reelle Funktion auf $T \times T$, so heisst d eine stetige Pseudometrik auf T, wenn $d(x,y) \geq 0$, $d(x,y) = d(y,x)$ und $d(x,z) \leq d(x,y) + d(y,z)$ für alle $x, y, z \in T$ gilt. Zur Vereinfachung der Sprechweise wollen wir für jede Unterkategorie $\mathcal{R} \subset \mathcal{K}$ vereinbaren :

Definition 1 : Eine invariante Pseudometrik auf $\mathcal{R}$ ist eine Vorschrift d, die jedem komplexen Raum $X \in \mathcal{R}$ eine stetige Pseudometrik d_X auf X so zuordnet, dass für jedes $Y \in \mathcal{R}$ und jede holomorphe Abbildung $f : X \rightarrow Y$ aus $\mathcal{R}$ gilt

$$d_Y(fx, fy) \leq d_X(x,y) \text{ für alle } x, y \in X .$$

Die Bezeichnung "invariant" erfährt eine gewisse Rechtfertigung dadurch, dass natürlich $d_Y(fx, fy) = d_X(x,y)$ gilt, wenn f biholomorph ist.

Die von CARATHEODORY definierte invariante Pseudometrik lässt sich unmittelbar auf die Kategorie $\mathcal{K}$ ausdehnen. Dazu gehen wir aus von der invarianten Metrik $D(a,b) = \left| \frac{a-b}{a\bar{b}-1} \right|$ im Einheitskreis $E := \{z \in \mathbb{C}: |z| < 1\}$ und setzen für alle $X \in \mathcal{K}$ und $x, y \in X$ (vergl. [5, 16])

$$C_X(x,y) := \sup_{f \in \mathcal{F}} D(fx, fy) ,$$

wobei $\mathcal{F} := \text{Hol}(X, E)$ die Menge aller holomorphen Funktionen f auf X sei mit $|f| < 1$. Damit gilt dann

Bemerkung 1 : (1) c ist eine invariante Pseudometrik auf $\mathcal{K}$;

(2) c_X ist eine Metrik genau dann, wenn $\mathcal{F}$ die Punkte von X trennt ;

(3) $c_X(x,y) = 1$ gilt genau dann, wenn x und y in verschiedenen

Zusammenhangskomponenten von X liegen ;

(4) Ist $c_X(x,y) < 1$, so gibt es ein $f \in \mathcal{F}$ mit $f(x) = 0$ und $|f(y)| = c_X(x,y)$.

Der Beweis ergibt sich leicht mit Hilfe des Satzes von MONTEL, der für beliebige komplexe Räume gilt, und mit Hilfe des Maximumprinzips für holomorphe Funktionen.

Wir wollen noch einige weitere Eigenschaften von c_X angeben und zeigen

Bemerkung 2 : Liegt für jedes $r < 1$ und jedes $x \in X$ die Kugel $K_r(x) := \{y \in X : c_X(x,y) < r\}$ relativ-kompakt in X, so ist X holomorph-konvex.

Beweis : Wir dürfen X als zusammenhängend voraussetzen. Es sei dann $K \subset X$ ein Kompaktum und $\hat{K} := \{y \in X : |f(y)| \leq \sup_{x \in K} |f(x)|$ für alle $f \in \mathcal{F}\}$ die $\mathcal{F}$-konvexe Hülle von K. K besitzt einen Durchmesser $d < 1$. Zu beliebig vorgegebenen Punkten $x \in K$ und $y \in \hat{K}$ gibt es stets ein $f \in \mathcal{F}$ mit $f(x) = 0$ und

$$c_X(x,y) = |f(y)| \leq \sup_{z \in K} |f(z)| \leq d.$$

Also kann $\hat{K}$ mit endlich vielen Kugeln mit einem Radius < 1 überdeckt werden, d.h. $\hat{K}$ ist kompakt. Also ist X $\mathcal{F}$-konvex und damit insbesondere holomorph-konvex.

Q.e.d.

Weiter gilt

Bemerkung 3 : Es seien $Y \subset X$ komplexe Räume mit der Eigenschaft, dass jede beschränkte holomorphe Funktion auf Y eindeutig zu einer holomorphen Funktion auf X fortgesetzt werden kann. Dann stimmen c_X und c_Y auf Y überein.

Beweis : Es sei $f : Y \to E$ eine beschränkte holomorphe Funktion auf Y und $\bar{f} : X \to C$ die eindeutig bestimmte Fortsetzung. Es genügt offenbar, $\bar{f}(X) \subset E$ zu beweisen. Das folgt jedoch mit einen bekannten Trick (vergl. [2]): Denn wäre $f(x_0) \notin E$ für ein x_0 aus X, so wäre im Widerspruch zur Voraussetzung die auf Y holomorphe Funktion $(f - f(x_o))^{-1}$ nicht auf ganz X holomorph fortsetzbar.

Q. e. d.

Wegen des Riemannschen Hebbarkeitssatzes (vergl. [9]) ist Bemerkung 3 z.B. stets dann anwendbar, wenn X ein normaler zusammenhängender komplexer Raum ist und Y ZARISKI-offen in X ist (d.h. wenn das Komplement von Y analytisch in X ist).

Wir wollen einen komplexen Raum X <u>beschränkt K-vollständig</u> nennen, wenn zu jeder nicht-diskreten Teilmenge $\Delta \subset X$ eine beschränkte holomorphe Funktion auf X existiert, die auf Δ nicht konstant ist. Jedes beschränkte Gebiet über dem C^n ist. z.B. beschränkt K-vollständig . Für alle $X \in \mathcal{K}$, $x \in X$ und alle r mit $0 < r \leq 1$ sei $Z_r(x)$ die x-Zusammenhangskomponente der Kugel $\{y \in X : c_X(x, y) < r\}$; für $r > 1$ werde $Z_r(x) := X$ gesetzt. Durch

$$\rho_X(x, y) := \inf \{r > 0 : x \in Z_r(y) \text{ und } y \in Z_r(x)\}$$

erhalten wir eine Pseudometrik ρ_X auf X (vergl. [13]) , und es gilt

<u>Satz 1 :</u> (1) <u>ρ ist eine invariante Pseudometrik auf $\mathcal{K}$;</u>

(2) <u>Ist X beschränkt K-vollständig, so ist ρ_X eine Metrik, die die Topologie von X erzeugt;</u>

(3) <u>Ist jede beschränkte holomorphe Funktion auf X konstant, so gilt $\rho_X \equiv 0$;</u>

(4) <u>Ist X normal und $Y \subset X$ eine ZARISKI-offene Teilmenge, so gilt $\rho_Y = \rho_X | Y$.</u>

<u>Beweis :</u> Die Invarianz von ρ folgt aus der Tatsache , dass stets $f(Z_r(x)) \subset Z_r(fx)$ gilt. Ist X beschränkt K-vollständig, so existiert zu jedem $x \in X$ ein $\varepsilon > 0$, so dass $Z_\varepsilon(x)$ relativ-kompakt in X liegt und $c_X(x, y) > 0$ für alle $y \in Z_\varepsilon(x)$ mit $y \neq x$ gilt. Daraus folgt , dass ρ_X die Topologie von X erzeugt. Es sei nun X normal und $Y \subset X$ ZARISKI-offen. Wir dürfen annehmen, dass Y jede Komponente von X schneidet, d.h. jede beschränkte holomorphe Funktion auf Y ist eindeutig auf X holomorph fortsetzbar. Wegen Bemerkung 3

gilt somit $c_X|Y = c_Y$ (was übrigens mit dem Maximumprinzip auch direkt folgt). Andererseits ist für jedes offene zusammenhängende $U \subset X$ auch die Menge $U \cap Y$ zusammenhängend, d.h. $\varrho_X|Y = \varrho_Y$.

Es seien $Y \subset X$ komplexe Räume und $f: Y \rightarrow Z$ eine holomorphe Abbildung. Notwendig dafür, dass f eine holomorphe Fortsetzung $\bar{f}: X \rightarrow Z$ gestattet, ist - falls Y dicht in X liegt - die folgende Bedingung:

(✱) Zu jedem $x \in X$ gibt es eine gegen x konvergente Punktfolge (y_n) in Y, so dass die Bildfolge (fy_n) in Z einen Häufungspunkt besitzt.

Im allgemeinen ist diese Bedingung nicht hinreichend, es gilt jedoch:

Satz 2: Es sei X ein normaler komplexer Raum und $Y \subset X$ eine ZARISKI-offene Teilmenge. Ist Z ein beschränkt K-vollständiger komplexer Raum und $f: Y \rightarrow Z$ eine holomorphe Abbildung, die die Bedingung (✱) erfüllt, so existiert für f genau eine holomorphe Fortsetzung $\bar{f}: X \rightarrow Z$.

Beweis: Sei x ein beliebiger Punkt in X. Wir wählen entsprechend (✱) eine Folge (y_n) in Y mit $x = \lim y_n$ und $q := \lim f(y_n) \in Z$. Sei $V \subset Z$ eine ε-Kugel (bez. ϱ_Z) um q so klein, dass $\bar{V}$ kompakt ist und eine Umgebung von $\bar{V}$ biholomorph äquivalent ist zu einer beschränkten lokal-analytischen Menge in einem C^m. Wegen Satz 1 existiert eine Umgebung U von x mit $f(U \cap Y) \subset V$, d.h. nach dem Riemannschen Fortsetzungssatz ist f in den Punkt x holomorph fortsetzbar. Diese Fortsetzung ist wegen (✱) natürlich eindeutig. Q.e.d.

Die Bedingung (✱) ist z.B. dann erfüllt, wenn die abgeschlossene Hülle $\overline{f(Y)} \subset Z$ vollständig bezüglich ϱ_Z ist. Denn die Folge (y_n) in (✱) ist wegen Satz 1 (4) eine CAUCHY-Folge bezüglich ϱ_Y, und wegen Satz 1 (1) ist dann auch die Bildfolge (fy_n) eine CAUCHY-Folge bezüglich ϱ_Z. Insgesamt erhalten wir daraus

Satz 3: Es seien $Y \subset X$ wie in Satz 2 und Z ein komplexer Raum, für den

ϱ_Z eine vollständige Metrik ist (z.B. wenn Z beschränkt K-vollständig und homogen ist). Dann kann jede holomorphe Abbildung $f: Y \to Z$ zu einer holomorphen Abbildung $\bar{f}: X \to Z$ fortgesetzt werden.

Dass ϱ_Z eine vollständige Metrik ist, wenn Z beschränkt K-vollständig und homogen (d.h. die Gruppe Aut (Z) aller Automorphismen von Z operiert transitiv auf Z; dann ist Z automatisch eine komplexe Mannigfaltigkeit) ist, kann folgendermassen eingesehen werden: Wegen Satz 1 ist ϱ_Z eine Metrik, die die Topologie von Z erzeugt. Es gibt also einen Punkt $z_o \in Z$ und ein $r > 0$, so dass die Kugel mit Radius r um z_o relativ-kompakt in Z liegt. Aus der Homogenität folgt sodann, dass jede Kugel vom Radius $< r$ relativ-kompkt in Z liegt, d.h. ϱ_Z ist insbesondere vollständig.

Es werde jetzt jedem komplexen Raum X eine weitere stetige Pseudometrik zugeordnet. Dazu betrachten wir die universelle Überlagerung $\tau: \tilde{X} \to X$ von X und setzen für alle $x, y \in X$

$$\sigma_X(x,y) := \varrho_{\tilde{X}}(\tau^{-1}(x), \tau^{-1}(y)) ;$$

dafür gilt dann

Satz 4: (1) σ ist eine invariante Pseudometrik auf $\mathfrak{X}$;

(2) Ist die universelle Überlagerung $\tilde{X}$ von X beschränkt K-vollständig, so ist σ_X eine Metrik, die die Topologie von X erzeugt;

(3) Ist jede beschränkte holomorphe Funktion auf $\tilde{X}$ konstant (z.B. wenn X eine zusammenhängende komplexe Liegruppe ist), so ist $\sigma_X \equiv 0$;

(4) $\sigma_X \geqslant \varrho_X \geqslant c_X$ und $\sigma_{\tilde{X}} = \varrho_{\tilde{X}}$;

(5) Für $X = E$ gilt $\sigma_E = \varrho_E = c_E = D$;

(6) Ist X normal und $Y \subset X$ eine ZARISKI-offene Teilmenge mit endlicher Fundamentalgruppe $\pi_1(Y)$, so ist eine Folge (y_n) in Y bereits dann eine CAUCHY-Folge bezüglich σ_Y, wenn sie in X konvergiert.

Beweis : Jede Decktransformation (d.i. ein Automorphismus g von $\tilde{X}$ mit $\tau = \tau g$) ist eine Isometrie bezüglich $\varrho_{\tilde{X}}$, und die Gruppe Γ aller Decktransformationen operiert eigentlich diskontinuierlich auf $\tilde{X}$. Deshalb gilt auch

(i) $$\sigma_X(x,y) = \inf_{\gamma \in \Gamma} \varrho_{\tilde{X}}(\tilde{x}, \gamma\tilde{y}),$$

wobei $\tilde{x} \in \tau^{-1}(x)$, $\tilde{y} \in \tau^{-1}(y)$ zwei beliebig gewählte Punkte sind, und aus (i) folgt, dass σ_X die Topologie von X erzeugt, wenn $\tilde{X}$ beschrankt K-vollständig ist. Die Invarianz von σ folgt einfach aus der entsprechenden Eigenschaft von ϱ und der Tatsache, dass zu jeder holomorphen Abbildung $f : X \to Y$ eine holomorphe Abbildung $\tilde{f} : \tilde{X} \to \tilde{Y}$ mit kommutativen Diagramm

$$\begin{array}{ccc} \tilde{X} & \xrightarrow{\tilde{f}} & \tilde{Y} \\ \tau\downarrow & & \downarrow \\ X & \xrightarrow[f]{} & Y \end{array}$$

existiert .

Ist X eine zusammenhängende komplexe Liegruppe, so ist auch $\tilde{X}$ eine komplexe Liegruppe, und wegen der Existenz genügend vieler einparametriger Untergruppen $C \to \tilde{X}$ ist jede beschränkte holomorphe Funktion auf $\tilde{X}$ konstant. (5) ist eine einfache Konsequenz des Schwarzschen Lemmas, und es fehlt nur noch der Nachweis von (6) : Die universelle Überlagerung $\omega : \tilde{Y} \to Y$ kann wegen $\pi_1(Y)$ endlich zu einer (i.a. verzweigten) analytischen Überlagerung $\varphi : \hat{X} \to X$ fortgesetzt werden, wobei $\hat{X}$ ein normaler komplexer Raum ist, der $\tilde{Y}$ als ZARISKI-offene Teilmenge enthält (vergl. [9]). Also stimmen $\sigma_{\tilde{Y}}$ und $\sigma_{\hat{X}}$ auf $\tilde{Y}$ überein. Ist nun (y_n) eine Folge in Y mit $\lim y_n \in X$, so kann eine Folge $(\tilde{y}_n)$ in $\tilde{Y}$ mit $\varphi\tilde{y}_n = y_n$ und $\lim \tilde{y}_n \in \hat{X}$ gefunden werden . Nun ist $\tilde{y}_n$ eine CAUCHY-Folge bezüglich $\sigma_{\hat{X}}$ und damit auch bezüglich $\sigma_{\tilde{Y}}$, d.h. die Bildfolge y_n ist eine CAUCHY-Folge bezüglich σ_Y.

Q.e.d.

Als Anwendung lässt sich zeigen

Satz. 5.: Es sei X ein normaler komplexer Raum und $N \subset X$ eine analytische Teilmenge, die folgender Bedingung genügt :

(**) Zu jedem $x \in N$ gibt es eine zusammenhängende Umgebung U von x in X, so dass die Fundamentalgruppe π_1 (U - N) endlich ist.

Dann gilt für $Y := X - N$ und jeden komplexen Raum Z, dessen universelle Überlagerung beschränkt K-vollständig ist: Jede holomorphe Abbildung $f: Y \to Z$, die die Bedingung (*) von Seite erfüllt, ist zu einer holomorphen Abbildung $X \to Z$ fortsetzbar.

Beweis: Wir dürfen $\pi_1(Y)$ endlich annehmen. Sei dann $x \in N$ und (y_n) eine Punktfolge in Y mit $\lim y_n = x$. Wegen Satz 4 (6) ist (y_n) eine CAUCHY-Folge bezüglich σ_Y, d.h. $(f(y_n))$ ist eine CAUCHY-Folge bezüglich der Metrik σ_Z. Wegen (*) durfen wir annehmen, dass die Folge $f(y_n)$ wenigstens einen Häufungspunkt besitzt, d.h. die Folge $f(y_n)$ ist bereits in Z konvergent.

Q.e.d.

Ist X eine komplexe Mannigfaltigkeit, so ist die Bedingung (**) für N z.B. dann erfüllt, wenn N eine Codimension > 1 in X besitzt. Nennen wir jetzt einen komplexen Raum X K-hyperbolisch, wenn σ_X eine vollständige Metrik auf X ist, so liefert der Beweis von Satz 5 speziell

Satz 5': Es seien X, Y und N wie in Satz 5. Ist Z ein K-hyperbolischer komplexer Raum, so ist jede holomorphe Abbildung $f: Y \longrightarrow Z$ zu einer holomorphen Abbildung $X \longrightarrow Z$ fortsetzbar.

Bevor wir auf die Notwendigkeit der Bedingung (**) eingehen, soll der Begriff "K-hyperbolisch" näher untersucht werden : Da jeder lokal-kompakte metrische Raum vollständig ist, auf dem eine Gruppe von Isometrien transitiv operiert (vergl. Satz 3), gilt zunächst :

Bemerkung 4 : Jede homogene komplexe Mannigfaltigkeit, deren universelle Überlagerung beschränkt K-vollständig ist, ist K-hyperbolisch.

Durch einfache Rechnung folgt weiter

Bemerkung 5 : Ein komplexer Raum X ist genau dann K-hyperbolisch wenn die universelle Überlagerung K-hyperbolisch ist.

Speziell ist also ein komplexer Raum K-hyperbolisch, wenn er Überlagerung eines kompakten Raumes ist und eine beschränkt K-vollständige Überlagerung besitzt.

Da die hyperbolischen Riemannschen Flächen gerade den homogenen Einheitskreis als universelle Überlagerung haben, sind also speziell alle hyperbolischen Riemannschen Flächen K-hyperbolisch, und der Begriff "K-hyperbolisch" erfährt dadurch eine gewisse Rechtfertigung. Weiter gilt

Bemerkung 6 : Sind X und Y K-hyperbolisch, so auch das direkte Produkt $X \times Y$ und jede analytische Teilmenge $A \subset X$.

Bemerkung 7 : Ist X ein K-hyperbolischer komplexer Raum und $N \subset X$ die Nullstellenmenge einer auf X beschränkten holomorphen Funktion, so ist auch das Komplement $X - N$ K-hyperbolisch.

Beweis : Es sei $Y := X - N$ und (y_n) eine CAUCHY-Folge in Y bezüglich σ_Y. Dann ist (y_n) auch eine CAUCHY-Folge bezüglich σ_X und konvergiert somit gegen ein $x \in X$. Nach Voraussetzung existiert eine holomorphe Abbildung $f : X \to E$ mit $f^{-1}(0) = N$. $E^* := \{t \in E : t \neq 0\}$ ist K-hyperbolisch. f bildet Y in E^* ab, d.h. $f(y_n)$ ist eine CAUCHY-Folge bezüglich σ_{E^*}, d.h. $f(x) \subset E^*$ und somit $x \in Y$. Also ist σ_Y vollständig.

Q.e.d.

Jedem komplexen Raum X ist in eindeutiger Weise die Normalisierung X^* zusammen mit einer diskreten eigentlichen holomorphen Abbildung $\varphi : X^* \to X$ zugeordnet (vergl. [9]). X^* ist ein normaler komplexer Raum, und man zeigt leicht

Bemerkung 8 : Mit X ist auch die Normalisierung X^* K-hyperbolisch.

Betrachten wir nun für jedes $n > 0$ das folgende ([1])

Beispiel : Es sei A eine kompakte K-hyperbolische komplexe Mannigfaltigkeit der Dimension n, die singularitätenfrei in einen komplex-projektiven Raum P_N eingebettet sei (ist z.B. A ein direktes Produkt von kompakten hyperbolischen Riemannschen Flächen, so ist das für geeignetes N stets erreichbar). A bestimmt im Vektorraum C^{N+1} einen analytischen Kegel X, dessen einzige Singularität der Nullpunkt $0 \subset C^{N+1}$ ist. $Y := X - \{0\}$ ist also eine komplexe Mannigfaltigkeit, und man hat eine holomorphe Abbildung $f : Y \to A$, die nicht holomorph auf X fortsetzbar ist, da $f(U \cap Y) = A$ für jede Umgebung U von 0 gilt. Die Bedingung (∗∗) ist also in Satz 5 wie in Satz 5' notwendig und kann nicht durch eine Bedingung an die Codimension von N ersetzt werden. Für komplexe Mannigfaltigkeiten ist die Bedingung (∗∗) jedoch vermutlich überflüssig (vergl. [10]).

Es sei X ein komplexer Raum, dessen universelle Überlagerung $\widetilde{X}$ beschränkt K-vollstandig ist. Dann ist die Gruppe Aut (X) aller Automorphismen von X als Gruppe aller biholomorphen Isometrien der Metrik σ_X eine reelle Liegruppe, die eigentlich auf X operiert (vergl. [13]). Ist $\Gamma \subset \mathrm{Aut}(\widetilde{X})$ die Gruppe aller Decktransformationen, so ist bekanntlich $\mathrm{Aut}(X) \approx N(\Gamma)/\Gamma$, wobei $N(\Gamma) = \{g \in \mathrm{Aut}(\widetilde{X}) : g\Gamma = \Gamma g\}$ der Normalisator von Γ in $\mathrm{Aut}(\widetilde{X})$ ist. Versehen wir jede Gruppe von Automorphismen mit der KO-Topologie (=Kompakt-Offen-Topologie), so gilt

Satz 6 : Es sei X ein zusammenhängender normaler K-hyperbolischer komplexer Raum und $Y \subset X$ das Komplement einer analytischen Teilmenge $N \subset X$, die die Bedingung (∗∗) aus Satz 5 erfüllt. Dann liefert die Zuordnung $g \longrightarrow g|Y$ einen topologischen Isomorphismus der Gruppe $\{g \in \mathrm{Aut}(X) : g(N) = N\}$ auf die Gruppe Aut(Y). Ist $N \neq \emptyset$, so kann Y nicht homogen sein, und ist N zusätzlich kompakt, so ist auch Aut(Y) kompakt. Ist N die Singularitätenmenge von X, so gilt Aut(X) = Aut(Y).

Beweis : Die Beschränkungsabbildung ψ von $G := \{g \in \mathrm{Aut}(X) : g(N) = N\}$ in Aut(Y) ist injektiv und stetig . Wegen Satz 5' ist ψ bijektiv . Da G und Aut (Y) Liegruppen mit abzählbarer Topologie sind, ist ψ ein topologischer Isomorphismus. Die folgenden Aussagen ergeben sich unmittelbar aus der Tatsache, dass G = Aut (Y) als Isometriengruppe eigentlich auf X operiert ([13]) bzw. dass die Singularitätenmenge von X invariant unter Aut(X) ist .

Daraus ergibt sich nun

Satz 7 : Für jeden kompakten K-hyperbolischen komplexen Raum X ist die Automorphismengruppe Aut(X) endlich.

Beweis : Nach einem Satz, der von BOCHNER und MONTGOMERY ([3]) für kompakte komplexe Mannigfaltigkeiten und von KERNER ([14]) für kompakte komplexe Räume bewiesen worden ist, ist Aut(X) eine komplexe Liegruppe, die holomorph auf X operiert. Ist G die 1-Komponente von Aut(X), so ist also speziell für jedes $x \in X$ die durch $g \longrightarrow gx$ definierte Abbildung $G \rightarrow X$ holomorph und wegen Satz 4 (3) konstant. Also ist $G = \{1\}$. Da X kompakt ist, ist Aut(X) als Isometriengruppe ebenfalls kompakt ([13]) und deshalb endlich.

Da auf einem kompakten komplexen Raum jedes holomorphe Vektorfeld integrierbar ist ([12]), ergibt sich insbesondere die

Folgerung : Ist X ein kompakter K-hyperbolischer komplexer Raum, so existiert ausser D = 0 kein holomorphes Vektorfeld auf X .

Wir wollen den Begriff "K-hyperbolisch" noch etwas erweitern und setzen

Definition : Ein zusammenhängender komplexer Raum X heisst hyperbolisch, wenn auf X eine vollständige stetige Metrik d existiert mit

$$d(fw, fz) \leqslant D(w, z)$$

für alle $w, z \in E$ und $f \in \mathrm{Hol}(E, X)$. Ein beliebiger komplexer Raum heisst hyperbolisch, wenn jede Zusammenhangskomponente hyperbolisch ist

Offensichtlich hängt diese Definition nicht ab von der speziellen

Wahl der invarianten Metrik D auf E. Betrachten wir nun auf jedem zusammenhängenden komplexen Raum Y die grösste Pseudometrik k_Y, so dass stets

$$k_Y(fw, fz) \leq D(w, z)$$

gilt. Diese existiert ; man setze nämlich für $x, y \in Y$

$$k_Y(x, y) := \inf \sum_{k=1}^{n} D(z_k, z_{k-1}),$$

wobei inf über alle endlichen Teilmengen $\{z_o, z_1, \ldots, z_n\}$ von E zu erstrecken ist, für die holomorphe Abbildungen $f_1, \ldots, f_n \in \mathrm{Hol}(E, Y)$ existieren mit $f_1(z_o) = x$, $f_n(z_n) = y$ und $f_k(z_k) = f_{k+1}$ für $1 \leq k < n$. Wir wollen k_Y auch die <u>KOBAYASHI-Pseudometrik</u> auf Y nennen (vergl. [15]) . Bezeichnen wir nun mit $\mathcal{K}_z$ die Kategorie aller zusammenhängenden komplexen Räume, so lässt sich leicht zeigen :

<u>Bemerkung 9</u> : <u>(1) k ist eine invariante Pseudometrik auf</u> $\mathcal{K}_z$;
<u>(2) X ist hyperbolisch genau dann , wenn k_X vollständig ist (denn mit d ist auch $k_X \geq d$ vollständig)</u>.

Daraus folgt nun unmittelbar für alle komplexen Räume X und Y

<u>Bemerkung 10</u> : <u>Ist Y hyperbolisch, so stimmen auf Hol(X, Y) die KO-Topologie und die Topologie der punktweisen Konvergenz überein.</u>

<u>Satz 8</u> : <u>Ist Y kompakt und hyperbolisch, so ist auch Hol(X, Y) kompakt .</u>

<u>Beweis</u> : Das direkte Produkt $Y^X = \prod_{x \in X} Y$ ist kompakt in der Produkttopologie; es kann aufgefasst werden als Menge aller Abbildungen von X in Y versehen mit der Topologie der punktweisen Konvergenz. Wegen Bemerkung 10 liegt Hol (X, Y) in Y^X abgeschlossen und damit kompakt (denn ein gleichmässiger Limes holomorpher Abbildungen ist wieder eine holomorphe Abbildung) .

Satz 8 ist Soezialfall des folgenden allgemeineren Satzes (vergl. [11]) :
<u>Ist X zusammenhängend und Y hyperbolisch, so ist die durch $(f, x) \rightarrow (fx, x)$ definierte Abbildung $\Phi : \mathrm{Hol}(X, Y) \times X \rightarrow Y \times X$ eigentlich (d.h. Φ ist stetig</u>

und $\Phi^{-1}(K)$ ist kompakt, wenn K kompakt ist). Dieser Satz kann als eine Art MONTELscher Satz für holomorphe Abbildungen gedeutet werden (vergl. [8]), denn für jedes Kompaktum $K = K_1 \times K_2 \subset Y \times X$ ist $\Phi^{-1}(K)$ genau dann kompakt, wenn die Menge $\{f \in \mathrm{Hol}(X,Y) : K_1 \cap f(K_2) \neq \emptyset\}$ kompakt ist. Zu jeder diskreten Punkfolge (f_k) in Hol(X, Y) gibt es also ein n_o mit

$$K_1 \cap f(K_2) = \emptyset \quad \text{für} \quad n > n_o$$

(d.h. die Folge (f_k) konvergiert gegen den idealen Rand von Y). Ist X zusätzlich kompakt, so ist nach DOUADY ([6]) Hol (X;Y) ein komplexer Raum und Φ eine holomorphe Abbildung. Daraus wird in [11] gefolgert (vergl. auch [4]):

Satz 9: Es sei X ein zusammenhängender kompakter komplexer Raum und Y ein kompakter komplexer Raum, der eine beschränkt separable Überlagerung besitzt. Dann gilt

(1) Es gibt nur endlich viele Zerlegungen von X, die durch holomorphe Abbildungen $f: X \to Y$ erzeugt werden (insbesondere gibt es nur endlich viele holomorphe Abbildungen von X auf Y mit zusammenhängenden Fasern).

(2) Zu vorgegebenen Punkten $x \in X$, $y \in Y$ gibt es nur endlich viele holomorphe Abbildungen $f : X \to Y$ mit $f(x) = y$,

(3) Zu jeder holomorphen Abbildung $f : Y \to Y$ gibt es eine natürliche Zahl $n > 0$, so dass die Iterierte $g = f^n$ eine Retraktion ist (d.h. $g^2 = g$). Insbesondere ist also jede surjektive holomorphe Abbildung $f : Y \to Y$ ein Automorphismus endlicher Ordnung von Y.

W. Kaup

LITERATUR

[1] ANDREOTTI, A. and W. STOLL: Extension of holomorphic maps, Ann. of Math. (2) 72, 312 - 349 (1960).

[2] BEHNKE, H. u. P. THULLEN: Theorie der Funktionen mehrerer komplexer Veränderlichen. Erg. d. Math. 3, Berlin: Springer 1934

[3] BOCHNER, S. a. D. MONTGOMERY: Groups on analytic manifolds. Ann. of Math. 48, 659-669 (1947).

[4] BOREL, A. a. R. NARASIMHAN: Uniqueness Conditions for Certain Holomorphic Mappings. Inventiones math. 2, 247-255(1967).

[5] CARATHEODORY, C.: ÜBER das Schwarzsche Lemma bei analytischen Funktionen von zwei komplexen Veränderlichen. Math. Ann. 97, 76-98 (1927).

[6] DOUADY, A.: Le problème des modules pour les sous-espaces analytiques compacts d'un espace analytique donné. Ann. Inst. Fourier 16, 1 - 95 (1966).

[7] GRAUERT, H.: Ein Theorem der analytischen Garbentheorie. Publ. Math. 5, 233-292 (1960).

[8] - , u. H. RECKZIEGEL: Hermitesche Metriken und normale Familien holomorpher Abbildungen. Math. Zeitschr. 89, 108 - 125 (1965)

[9] - , u. R. REMMERT: Komplexe Räume. Math. Ann. 136, 245-318 (1958)

[10] HUBER, H.: Über analytische Abbildungen Riemannscher Flächen in sich. Comment. Math. Helv. 27, 1-72 (1953)

[11] KAUP, W: Endlichkeitssätze für Systeme holomorpher Abbildungen in hyperbolische Räume. In Vorbereitung

[12] KAUP, W.: Infinitesimale Transformationsgruppen komplexer Räume. Math. Ann. 160, 72 - 92 (1965).

[13] KAUP, W.: Reelle Transformationsgruppen und invariante Metriken auf komplexen Räumen. Inventiones math. 3, 43 - 70 (1967).

[14] KERNER, H.: Uber die Automorphismengruppen kompakter komplexer Räume. Arch. Math. 11, 282 - 288 (1960).

[15] KOBAYASHI, S.: Intrinsic metrics on complex manifolds. Bull. Amer. Math. Soc. 73, 347-349 (1967).

[16] REIFFEN, H. J.: Die Carathéodorysche Distanz und ihre zugehorige Differentialmetrik. Math. Ann. 161, 315-324 (1965)

[17] REMMERT, R.: Holomorphe und meromorphe Abbildungen komplexer Räume. Math. Ann. 133, 328-370 (1957)

CENTRO INTERNAZIONALE MATEMATICO ESTIVO

(C. I. M. E.)

A. KORANYI

"HOLOMORPHIC AND HARMONIC FUNCTIONS ON BOUNDED SYMMETRIC DOMAINS"

Corso tenuto ad Urbino dal 5 al 13 luglio 1967

HOLOMORPHIC AND HARMONIC FUNCTIONS ON BOUNDED SYMMETRIC DOMAINS

by

A. KORÁNYI
Yeshiva University (New York)

The main purpose of these lectures is to study questions of elementary analysis on bounded symmetric domains, namely the realization of these domains as generalizations of the unit disc and the upper halfplane, the study of the structure of their boundary and the boundary behaviour of holomorphic functions. This is done in sections 3 to 6 which contain material otherwise available only in journals (mainly [14], [16] and [24]). Some slight simplifications and improvements have been made here; it will, by the way, be apparent that the subject still has plenty of open problems.

A second purpose is to make all this more accessible to the analyst who is not an expert in Riemannian geometry. Our study uses parts of the theory of Riemannian symmetric spaces. For this standard treatise is that of Helgason [8]; another very clear and concise introduction can be found in chapters 1, 2, 8 of the book of Wolf [22]. These books contain all that is needed here, but they also contain much more. In our section I we outline how the part of the theory actually needed can be built up with minimum effort. In particular, we do not need locally symmetric spaces and we can avoid most of the topological difficulties. In this section most proofs are omitted, but they should be easy to fill in with the help of the references.

Section 2 is a slight rearrangement of material contained in [8].

Because it is the basis of all that comes later, we present this material with proofs. (This does not mean all proofs; here and also later some the proofs, preferably the lengthy and uninstructive ones, will be omitted. In these cases easily traceable references will be given).

The prerequisites for the reading of these notes are the following.

(i) The fundamental facts about Lie groups and their homogeneous spaces ([8] Ch. II, or [5] Ch. IV, or , in a nutshell, [22] Ch. I sec. 5). For this, in turn, one needs the basic definitions about differentiable manifolds (same references) and some facts about covering spaces and covering groups ([22] Ch. I sec. 8, or [5] Ch. II; a good reference for all the above is also [11]).

(ii) The fundamentals of the theory of semisimple Lie groups and Lie algebras. (Best exposition for our purpose in [8] Ch. III. Also to be found in [11] , [13] , etc.)

(iii) Some of the elementary facts of Riemannian geometry: The existence of geodesics, normal neighborhoods, the notion of completeness. (A few pages in any book on the subject; also in [8] and [22].).

A. Koranyi

I. Preliminaries.

Generalities on complex domains. By a domain we mean a connected open subset D of C^n. If D' is another domain and $f : D \longrightarrow D'$ is holomorphic with holomorphic inverse, we say that f is an isomorphism. If $D' = D$, f is called an automorphism of D. The automorphisms of D form a group, denoted $G(D)$. D is said to be homogeneous if $G(D)$ is transitive on D.

If D is bounded, or isomorphic with a bounded domain, then D has an invariant Hermitian metric, the Bergman metric. More precisely, to every $z \in D$ there is attached a positive definite Hermitian matrix $\{ g_{jk}(z) \}$ $(j, k = 1, \ldots, n)$; the length of a tangent vector

$$v = \sum (a_j \frac{\partial}{\partial z_j} + \bar{a}_j \frac{\partial}{\partial \bar{z}_j}$$

is defined by $|v|_D^2 = \sum g_{jk}(z)\, a_j \bar{a}_k$ (every real tangent vector can be written in this form); if $f : D \longrightarrow D'$ is an isomorphism, then it is also an isometry, i. e. $|(df)v|_{D'} = |v|_D$. (For the proofs see e. g. [8] p. 293-300).

We denote by $I(D)$ the group of all isometries of D as a Riemannian space, and give $I(D)$ the compact-open topology. By the Mayers-Steenrod theorem $I(D)$ is a Lie group (a proof in a special case, still sufficiently general for our later use, is in [8] p. 170-172). By the theorem of Weierstrass, G(D) is a closed subgroup of I(D) ; in particular it is a Lie group.

A. Koranyi

Symmetric domains and symmetric spaces

Definition : A bounded domain D is symmetric if for every $z \in D$ there exists $s_z \in G(D)$ such that (i) $s_z^{-1} = s_z$ and (ii) z is an isolated fixed point of s_z.

If we equip D with its Bergman metric, it becomes a member of the following class of spaces:

Definition : A connected Riemannian space M is symmetric if for every $p \in M$ there exists an isometry s_p of M onto itself such that $s_p^{-1} = s_p$ and p is an isolated fixed point of s_p.

Given a complex manifold L of n (complex) dimensions we can talk about its underlying real manifold M (2n-dimensional). Using a local coordinate system $z_1, \ldots, z_n$ at a point $p \in L$, every (real) tangent vector at p can be written in the form

$$v = \sum \left(a_j \frac{\partial}{\partial z_j} + \bar{a}_j \frac{\partial}{\partial \bar{z}_j}\right).$$

The mapping $a_j \rightarrow ia_j$ $(j = 1, \ldots, n)$ induces a (real) linear transformation J_p on M_p (for each p) which is easily seen to be independent of the coordinate system and is such that $J_p^2 = -I$. In this situation we say that J is a complex structure on M; L is reconstructible from M and J. If M' is another manifold with complex structure J', it is immediate that a differentiable map $f : M \rightarrow M'$ is holomorphic if and only if $df \circ J = J' \circ df$

at every point of M (Cauchy - Riemann equations).

We say that M is a <u>Hermitian space</u> if M is a Riemannian space with a complex structure J such that $|Jv| = |v|$ for every vector v.

Clearly every domain with Bergman metric is a Hermitian space; every bounded symmetric domain is a Hermitian symmetric space, a notion defined as follows.

<u>Definition</u> : A <u>Hermitian symmetric space</u> is a Hermitian space M such that for every $p \in M$ there exists a <u>holomorphic</u> isometry s_p such that $s_p^{-1} = s_p$ and p is an isolated fixed point of s_p.

In order to study bounded symmetric domains one looks at Riemannian symmetric spaces first, then singles out the Hermitian ones, and finally the bounded domains.

<u>Riemannian symmetric spaces</u> . Let M be Riemannian symmetric, let $p \in M$, then $(ds_p)_p$ is an involutive linear transformation on M_p; since p is an isolated fixed point, $(ds_p)_p = -I$. Hence if γ is a geodesic with $\gamma(0) = p$, then $s_p \gamma(t) = \gamma(-t)$ (for small t, at the moment). It is immediate that $T_t = s_{\gamma(t/2)} s_{\gamma(0)}$ is a one-parameter (local) group of isometries, called <u>transvections</u> based at p, which has γ as one of its orbits, (and induces parallel translation of vectors along γ). Now a simple process of "continuation" shows that every geodesic can be continued indefinitely, so M is complete.

A. Koranyi

Let $p, q \in M$. By completeness there is a geodesic segment joining them; let r be its midpoint. Then $g = s_q s_r$ carries p to q and is on a one-parameter group of transvections. This shows that $I_o(M)$ (the identity component of $I(M)$) is transitive on M.

(These remarks show that Riemannian or Hermitian symmetric spaces and bounded symmetric domains may be defined in new equivalent ways. We formulate the new definition here for domains:

A bounded domain D is symmetric if and only if it is homogeneous and for <u>some</u> point $z_o \in D$ there exists $s \in G(D)$ such that $s^{-1} = s$ and z_o is an isolated fixed point of s.)

Let M be Riemannian symmetric and let $G = I_o(M)$. We fix a point $o \in M$, and denote by K the isotropy group at o (i.e. $K = \{ g \in G \mid g \cdot o = o \}$). K is compact, contains no normal subgroup of G, and M can be identified with the coset space G/K.

$g \longrightarrow s_o g s_o$ is an involutive automorphism of G; it is easy to see that its fixed point set has the same identity component as K. It induces an involutive automorphism σ of the Lie algebra $\mathfrak{g}$ of G. The fixed point set of σ is $\mathfrak{k}$, the Lie algebra of K. Denoting by $\mathfrak{p}$ the (-1)-eigenspace of σ, we have $\mathfrak{g} = \mathfrak{k} + \mathfrak{p}$ as a vector space direct sum. It is now not hard to check that the one-parameter groups of transvections based at o are exactly the $\{ \exp tY \}$ with $Y \in \mathfrak{p}$. Denoting $P = \{ \exp Y \mid Y \in \mathfrak{p} \}$ it follows that $G = KP = PK$.

It is also easy to see that $\psi : \mathfrak{p} \to M_o$ defined by $\psi(Y) = \frac{d}{dt} (\exp tY) \cdot o|_{t=0}$ is a vector space isomorphism commuting with the action of K (i.e. $\psi \circ \mathrm{ad}(k)|_{\mathfrak{p}} = (dk)_o \circ \psi$ for all $k \in K$). Transporting the Riemannian structure from M_o to $\mathfrak{p}$ by ψ, we get an ad(K) - invariant positive definite quadratic form Q on $\mathfrak{p}$.

Definition : An ortogonal involutive Lie algebra (oiLa) is a triple $(\mathfrak{g}, \sigma, Q)$ such that

(i) $\mathfrak{g}$ is a real Lie algebra

(ii) σ is an involutive automorphism of $\mathfrak{g}$. (We shall always denote its eigenspaces for 1 and -1 by $\mathfrak{k}$ and $\mathfrak{p}$; we then have $[\mathfrak{k}, \mathfrak{k}] \subset \mathfrak{k}$, $[\mathfrak{k}, \mathfrak{p}] \subset \mathfrak{p}$, $[\mathfrak{p}, \mathfrak{p}] \subset \mathfrak{k}$).

(iii) $\mathfrak{k}$ contains no non-zero ideal of $\mathfrak{g}$

(iv) Q is an $\mathrm{ad}(\mathfrak{k})$-invariant positive definite quadratic form on $\mathfrak{p}$.

We have associated an oiLa to every Riemannian symmetric space. Conversely, given any oiLa $(\mathfrak{g}, \sigma, Q)$ we can always reconstruct a simply connected Riemannian symmetric space from it by taking the simply connected group $\tilde{G}$ with Lie algebra $\mathfrak{g}$, the analytic subgroup $\tilde{K}$ corresponding to $\mathfrak{k}$, and using Q to define a Riemannian metric on $\tilde{G}/\tilde{K} = M$. (It may be that $\tilde{G} \neq I_o(M)$, but this causes no difficulty. Details are e.g. in [22] p. 242). In general there are several symmetric spaces corresponding to an oiLa (since non-connected groups may sometimes be used instead of $\tilde{K}$), but in the cases that interest us it will turn out that the spaces are automatically simply connected.

A. Koranyi

Orthogonal involutive Lie algebras. The oiLa $(\mathfrak{g}, \sigma, Q)$ is called Euclidean if $[\mathfrak{p}, \mathfrak{p}] = 0$ (the corresponding simply connected space is Euclidean). It is called irreducible if it is not Euclidean and $\mathrm{ad}(\mathfrak{k})$ acts irreducibly on $\mathfrak{p}$.

By rather simple algebraic reasoning (e.g. [22] p. 235-237) the following decomposition theorem can be proved:

Given any oiLa $(\mathfrak{g}, \sigma, Q)$, $\mathfrak{g}$ is a direct sum $\mathfrak{g} = \mathfrak{g}_0 \oplus \cdots \oplus \mathfrak{g}_t$ where each $(\mathfrak{g}_j, \sigma|\mathfrak{g}_j, Q|\mathfrak{g}_j)$ is an oiLa, Euclidean for $j = 0$, irreducible for $j > 0$. For $j > 0$, $\mathfrak{g}_j$ is semisimple, $\mathfrak{k}_j = [\mathfrak{p}_j, \mathfrak{p}_j]$, and $Q|\mathfrak{p}_j$ is a scalar multiple of the restriction of the Killing form to $\mathfrak{p}_j$.

Let $(\mathfrak{g}, \sigma, Q)$ be an irreducible oiLa. One defines its dual $(\mathfrak{g}^*, \sigma^*, Q^*)$ by $\mathfrak{g}^* = \mathfrak{k} + i\mathfrak{p}$ and by transporting σ and Q to $\mathfrak{g}^*$ in an obvious way. It is clear that one of $\mathfrak{g}$, $\mathfrak{g}^*$ is compact, the other non-compact.

In general, if the decomposition of $(\mathfrak{g}, \sigma, Q)$ contains no Euclidean factor and each factor is compact (resp. non-compact), we say that it is of compact (respt. non compact) type.

If G/K is Riemannian symmetric and the corresponding oiLa is of non-compact type, then G/K must be simply connected; this is shown by a straightforward argument in [8].

In the irriducible compact case, $\mathfrak{g}$ may be simple or non-simple. In the latter case it is easy to show ([22]; p. 238) that $\mathfrak{g} = 1 \oplus 1$, with 1 compact simple, and $\mathfrak{k} \cong 1$ is the diagonal subgroup. The dual of such a $\mathfrak{g}$ is the corresponding compler simple algebra. If $\mathfrak{g}$ is compact simple, then σ may be any involutive automorphism of it. (The classification of the latter amounts exactly to the classification of all real simple Lie algebras (e.g. 22 , p. 238); this remark is irrelevant for our purposes since in the Hermitian case the classification is considerably easier).

OiLa's of bounded symmetric domains. First we have to single out the oiLa's which are associated to Hermitian symmetric spaces. If M is Hermitian symmetric, M_0 has a complex structure operator J_0 commuting with the action of K . ψ carries J_0 to a complex structure J on $\mathcal{C}$, commuting with $\mathrm{ad}\,\mathfrak{k}$. It is easy to see that each $\mathcal{C}_j$ in the decomposition into Euclidean an irreducible factors of $\mathfrak{g}$ is invariant under J .

Now we consider only the irreducible case. By Schur's lemma $\mathrm{ad}\,\mathfrak{k}$ extended to $\mathcal{C}^{\mathbb{C}}$ (the complexification of $\mathcal{C}$) is not irreducible. By a property of semisimple algebras (e.g. [13] p. 223) this implies that $\mathfrak{k}$ is not semisimple. It follows immediately that $\mathfrak{g}$ must be simple .

$\mathfrak{k}$ is still reductive (e.g. the Killing form of $\mathfrak{g}$ can be seen to be negative definite on $\mathfrak{k}$), so $\mathfrak{k} = \mathfrak{z} \oplus [\mathfrak{k}, \mathfrak{k}]$ where $\mathfrak{z}$ is the center and $[\mathfrak{k},\mathfrak{k}]$ is semisimple. By irreducibility, $\mathfrak{z}$ is one-dimensional and the corresponding analytic subgroup of $\mathrm{ad}(K)|_{\mathcal{C}}$ acts as $\{(\cos\theta)I + (\sin\theta)J\}$. In particular, $J \in \mathrm{ad}(\mathfrak{z})|_{\mathfrak{k}}$ and $J \in \mathrm{ad}(\exp \mathfrak{z})|_{\mathfrak{k}}$.

Furthermore, still in the irreducible case, one sees

that $\mathfrak{k}$ is the centralizer of $\mathfrak{z}$ in $\mathfrak{g}$, since $\mathfrak{g}$ has no center and since $\mathfrak{k}$ is a maximal proper subalgebra (this follows since ad $\mathfrak{k}$ acts irreducibly on $\mathfrak{p}$).

Let now $M = G/K$, $G = I_0(M)$ be Hermitian symmetric of compact type. Then K is contained in the centralizer of exp $\mathfrak{z}$. But the centralizer of a torus in a compact connected group is always connected, so K is connected. It follows that K contains a maximal torus, hence also the center of G ; hence the center of G is trivial. By some further argument it follows that M is simply connected [8] p. 214-216.

Suppose now that D is a bounded symmetric domain with the Bergman metric. Let $\tilde{D}$ be its universal covering space; it is still Hermitian symmetric . So we have $\tilde{D} = M_0 x M_1 x$ $x \ldots x M_t$ with M_0 Euclidean, the other M_j irreducible Hermitian symmetric. On D we have the bounded holomorphic functions $z_1, \ldots, z_n$; their differentials span the cotangent space at any point. These functions lift to $\tilde{D}$ and have still the same properties. So they can not all be constant on any of the M_j. Now Liouville's theorem shows that M_0 does not occur, and the maximum principle shows that all the M_j's are non compact. So D is of non-compact type, in particular D is automatically simply connected.

Cartan subalgebras. Let $M = G/K$, $G = I_0(M)$ be any Riemannian symmetric space; we have the corresponding decomposition $\mathfrak{g} = \mathfrak{k} + \mathfrak{p}$. Every subalgebra of $\mathfrak{g}$ contained in $\mathfrak{p}$ is abelian, by $[\mathfrak{p}, \mathfrak{p}] \subset \mathfrak{k}$. A maximal such subalgebra $\mathfrak{a}$ is called a Cartan subalgebra of the pair $(\mathfrak{g}, \mathfrak{k})$.

Similary as in the case of Cartan subalgebras of Lie algebras one can prove (e. g. [22] p. 252-253) that

(i) every $\mathfrak{a}$ contains elements X (called regular elements) such that $\mathfrak{a}$ equals the centralizer of X in $\mathfrak{p}$,

(ii) if $\mathfrak{a}'$ is another Cartan subalgebra, then there exists $k \in K$ such that $\mathrm{ad}(k)\mathfrak{a} = \mathfrak{a}'$;

(iii) $\mathfrak{p} = \bigcup_{k \in K} \mathrm{ad}(k)\,\mathfrak{a}$ (immediate from (ii) and Zorn's lemma).

From (iii) it follows that $P = \exp \mathfrak{p} \subset K A K$, where $A = \exp \mathfrak{a}$. Using that $G = P K$, it follows that $G = K A K$, a fundamental fact. (E. g. if G/K is the 2-sphere, this amounts to decomposing an arbitrary rotation in terms of the Euler-angles).

$\mathrm{ad}(\mathfrak{a}^{\mathbb{C}})$ is a commutative algebra of semisimple linear transformations on $\mathfrak{g}^{\mathbb{C}}$, so one can introduce roots and root spaces with respect to it. It is known that one obtains in this way one of the standard root systems (a proofs is in the Appendix of [19]). The Weyl group of this root system (called the small Weyl group) is realized by the subgroup of K which normalizes $\mathfrak{a}$, if the corresponding symmetric space is realized as G/K ([8] p. 246). These last remarks will not be used very much, a large part of what we do is independent of them.

A. Koranyi

2. Imbedding Theorems of Borel and Harish-Chandra.

We start now with an irreducible oiLa $(\mathfrak{g}, \sigma, Q)$, $\mathfrak{g} = \mathfrak{k} + \mathfrak{p}$ such that $\mathfrak{k}$ is not semisimple, and we want to show that there exists a corresponding bounded symmetric domain. (Note that we do not know yet whether there is a corresponding Hermitian symmetric space; we surely can construct a Riemannian symmetric G/K and carry over the complex structure J of $\mathfrak{p}$ to the tangent space at every point, but it is not clear yet that the "almost complex structure" obtained in this way really derives from a complex structure. This question will be settled without using the Newlander-Nirenberg integrability condition).

Let $\mathfrak{g}^c$ be the complexification of $\mathfrak{g}$. extends to $\mathfrak{g}^c$, we have $\mathfrak{g}^c = \mathfrak{k}^c + \mathfrak{p}^c$. Let $\mathfrak{g}_u = \mathfrak{k} + i\mathfrak{p}$, let τ be the conjugation of $\mathfrak{g}^c$ with respect to $\mathfrak{g}_u$. We know that $\mathfrak{k}$ has a one-dimensional center $\mathfrak{z}$ spanned by an element Z such that $J^2 = -I$ for $J = ad(Z)$.

Let $\mathfrak{h}$ be any maximal abelian subalgebra of K. Each ad(H) $(H \in \mathfrak{h})$ is skew-Hermitian with respect to the positive definite Hermitian from $B_\tau(X, Y) = -B(X, \tau Y)$ on $\mathfrak{g}^c$ (trivial checking). It follows that $\mathfrak{h}^c$ is a Cartan subalgebra of $\mathfrak{g}^c$. With respect to this we take a standard basis $H_\alpha, \ldots, E_\alpha; \ldots$ such that $[E_\alpha, E_{-\alpha}] = H_\alpha$ and fixing the lengths of the E_α so as to satisfy $\tau E_\alpha = -E_{-\alpha}$ for $\alpha \in \Phi$, which is defined as follows.

Each ad(H) $(H \in \mathfrak{h})$ preserves $\mathfrak{k}^c$ and $\mathfrak{p}^c$ since $\mathfrak{h} \in \mathfrak{k}$ and $[\mathfrak{k}, \mathfrak{k}] \subset \mathfrak{k}$, $[\mathfrak{k}, \mathfrak{p}] \subset \mathfrak{p}$. Hence for every root α, either $E_\alpha \in \mathfrak{k}^c$ or $E_\alpha \in \mathfrak{p}^c$. In the first case α is called a <u>compact</u>, in the second a <u>non-compact</u> root. We denote by Φ the set of positive non-compact roots. We define

$$\mathfrak{p}^+ = \sum_{\alpha \in \Phi} \mathbb{C} E_\alpha$$

$$\mathfrak{p}^- = \sum_{\alpha \in \Phi} \mathbb{C} E_{-\alpha}$$

Then $\mathfrak{p}^c = \mathfrak{p}^+ + \mathfrak{p}^-$ (vector space direct sum).
It is now clear that $\tau E_\alpha = E_{-\alpha}$ ($\alpha \in \Phi$) can be arranged.

Since $\mathfrak{h} \subset \mathfrak{g}_u$ we have $H_\alpha \in i\mathfrak{h}$ for each α. $\mathcal{Z} \in \mathfrak{h}$ since $\mathfrak{k}$ is the centralizer of $\mathcal{Z}$. We order the dual of $i\mathfrak{h}$ so that $-i\,\alpha(Z) > 0$ implies $\alpha > 0$ (this is trivially possible). $J = ad(Z)$ extends to a complex linear transformation on $\mathfrak{g}^c$. For $\alpha \in \Phi$ we have $JE_\alpha = \alpha(Z) E_\alpha$; by $J^2 = -I$, $\alpha(Z) = \pm i$. By the choice of ordering, $\alpha(Z) = i$. Similarly $JE_{-\alpha} = -iE_{-\alpha}$.

We also introduce a basis for $\mathfrak{p}$, defining, for all $\alpha \in \Phi$,

$$X = E_\alpha + E_{-\alpha}$$

$$Y = i(E_\alpha - E_{-\alpha})$$

and we have $JX_\alpha = Y_\alpha$, $JY_\alpha = -X_\alpha$

<u>Lemma 1.</u> $\mathfrak{p}^+$ and $\mathfrak{p}^-$ are abelian subalgebras of $\mathfrak{g}^c$, stable under $ad\,\mathfrak{k}^c$.

Proof. $ad(Z)$ has eigenvalues $\pm i$, 0 on $\mathfrak{g}^c$ with corresponding eigenspaces $\mathfrak{p}^{\pm}$, $\mathfrak{k}^c$. By a standard argument (the Jacobi identity), if X, Y are eigenvectors for the eigenvalues

λ, μ, then $[X, Y]$ is either 0 or an eigenvector for $\lambda+\mu$. Hence $[\mathcal{C}^+, \mathcal{C}^+] = [\mathcal{C}^-, \mathcal{C}^-] = 0$, $[\mathcal{K}^{\mathbb{C}}, \mathcal{C}^{\pm}] \subset \mathcal{C}^{\pm}$.

Now let G^c be the simply connected group with Lie algebra $\mathcal{G}^c$, and let G, $G_{\mathcal{U}}$, K, K^c, P^+, P^- be the analytic subgroups for the corresponding subalgebras of $\mathcal{G}^c$.

<u>LEMMA 2</u>. $G_{\mathcal{U}}$, G, K^c, K, $P^{\pm}$ are closed subgroups of G^c. $\exp: \mathcal{C}^{\pm} \to P^{\pm}$ are homeomorphisms.

Proof. $\mathcal{T}$ induces an involutive automorphism $\mathcal{T}^*$ of G^c. $G_{\mathcal{U}}$ is the identity component of the fixed point set of $\mathcal{T}^*$, hence closed. The same argument with $\mathfrak{S}$ and conjugation with respect to $\mathcal{G}$ works for K^c and G. K is closed for it is the identity component of $K^c \cap G$.

$ad(\mathcal{C}^+)$ as a subalgebra of $ad(\mathcal{G}^c)$ is an algebra of nilpotent linear transformations, and it is a faithful representation of $\mathcal{C}^+$. Exponentiation maps $ad\,\mathcal{C}^+$ homeomorphically onto the closed subgroup $ad_{G^c}(P^+)$ of the adjoint group of G^c. Now P^+, being the preimage under ad is a covering of the simply connected group $ad_{G^c}(P^+)$. Hence $\exp: \mathcal{C}^+ \to P^+$ must be a homeomorphism; also P^+ is closed.

<u>LEMMA 3</u>. $K^c P^-$ is a closed analytic subgroup of G^c with Lie algebra $\mathcal{K}^c + \mathcal{C}^-$.

Proof. $K^c P^-$ is a group since $\mathcal{C}^-$ is $ad(\mathcal{K}^c)$-invariant and $P^- = \exp \mathcal{C}^-$. $\mathcal{K}^c + \mathcal{C}^-$ is the normalizer of $\mathcal{C}^-$ in $\mathcal{G}^c$ (trivial). The normalizer of $\mathcal{C}^-$ in G^c is a closed subgroup with Lie algebra $\mathcal{K}^c + \mathcal{C}^-$. The identity component of this must be $K^c \cdot P^-$.

<u>REMARK</u>. Looking at the adjoint group of G^c one sees that $K^c \cap P^- = \{e\}$. Hence $K^c \cdot P^-$ is a semi-direct product.

We denote the homogeneous space $G^c / K^c P^-$ by M^* (the reason for this notation will be clear soon). For brevity we denote the identity coset by x. Since $\mathfrak{k}^c + \mathfrak{p}^-$ is a complex subalgebra, M^* has a natural complex structure and G^c acts on M^* by holomorphic maps (standard fact, see e.g. [22] p. 258). The tangent space M_x is isomorphic with $\mathfrak{p}^+$ as a complex vector space.

In the following three theorems we examine the orbit of x in M^* under G_U, G, and P^+.

THEOREM 1. G_U/K is a compact Hermitian symmetric space. $i : G_U/K \to M^*$ defined by $i(gK) = gx$ is a G_U-equivariant holomorphic diffeomorphism onto M^*.

<u>Proof</u>. $\mathfrak{g}_U \cap (\mathfrak{k}^c + \mathfrak{p}^-) = \mathfrak{k}$, which shows that i is a homeomorphism of an open neighborhood of eK onto an open set. By G_U -equivariance, which is clear, this is true at every gK , hence the image of G_U/K is open.

G_U/K is compact, since G_U is. Then its image is also compact, hence closed. By connectedness, i maps onto M^*.

By the remarks above, i is a covering. But M^* is simply connected (since $G^{\mathbb{C}}$ simply connected and $K^c \cdot P^-$ connected), so i is a diffeomorphism. i^{-1} carries the complex structure of M^* back to G_U/K, and the action of G_U becomes holomorphic by equivariance. So G_U/K is Hermitian symmetric.

THEOREM 2. $M = G/K$ is a Hermitian symmetric space. The map $j : M \to M^*$ defined by $j(gK) = gx$ is a G-equivariant holomorphic diffeomorphism onto an open subset of M^*.

Proof. G-equivariance is trivial. Now to show one-to-one it is enough to show $G \cap K^c P^- = K$. Since $G = KP (P = \exp \mathcal{P})$, for this is enough to show $P \cap K^c P^- = \{e\}$.

Suppose $p = kp^-$ ($p \in P$, $k \in K^c$, $p^- \in P^-$). Let σ^*, τ^* be the involutive automorphisms of G^c induced by σ, τ. We have

$$p^{-1} = \sigma^* \tau^* p = \sigma^* \tau^* (k\, p^-) = k\, (p^-)^{-1}$$

whence $p = p^- k^{-1}$, $p^2 = (p^-)^2$. Applying τ^* to this, $p^{-2} = \tau^* (p^-)^2 \in P^+$, and therefore $p^2 = e$ (since Ad_{G^c} is a faithful representation of $P^{\pm}$ onto upper resp. lower triangular matrices), and hence $p = e$.

j is a diffeomorphism onto an open set for the same reason as i was in theorem 1. Also similarly to i^{-1}, j^{-1} brings a complex structure to M with respect to which G acts holomorphically ; so M is Hermitian symmetric. It is also easy to see that this complex structure induces J on $\mathcal{P}$.

In the following we do not write out i and j, but consider $G_u / K = M^*$ and $M \in M^*$.

Note that, in the special case of $G^c = SL(2, C)$, what we have done so far amounts to having constructed the Riemann sphere S^2 and imbedded the unit disc into it as the lower hemisphere. Next we imbed the complex plane into S^2 by stereographic projection.

THEOREM 3. $\xi : \mathcal{P}^+ \to M^*$ defined by $\xi(X) = (\exp X)x$ is a K^c-equivariant holomorphic diffeomorphism of the complex vector space $\mathcal{C}^+$ onto an open subset of M^*.

Proof To prove that ξ is one-to-one we must show $P^+ \cap K^c P^- = \{e\}$. Assume that $e \neq g \in P^+ \cap K^c P^-$. Let $X \in \mathcal{C}^+$ such that $\exp X = g$. Then $X = \sum c_\beta E_\beta \; (\beta \in \Phi)$. Let β be the lowest root such that $c_\beta \neq 0$. Denoting $\mathcal{N}^+ = \sum_{\alpha > 0} \mathbb{C} E_\alpha$, we have $[X, E_{-\beta}] \equiv c_\beta H_\beta \pmod{\mathcal{N}^+}$ and, exponentiating, $\mathrm{ad}(g) E_{-\beta} \equiv E_{-\beta} + c_\beta H_\beta \pmod{\mathcal{N}^+}$. But K^c P^- normalizes $\mathcal{C}^-$ so we must have $\mathrm{ad}(g) E_{-\beta} \in \mathcal{C}^-$, which is a contradiction.

$\exp : \mathcal{C}^+ \to P^+$ is a diffeomorphism. $p^+ \mapsto p^+ \cdot x$ is everywhere regular, since $\mathcal{G}^c = \mathcal{C}^+ + (\mathcal{R}^c + \mathcal{C}^-)$. Its image is open by a dimension count.

Finally, let $k \in K^c$. We have $\xi\,(\mathrm{ad}(k)\,X) = \exp\,(\mathrm{ad}(k) X)\, x = k (\exp X)\, k^{-1} x = k\,(\exp X)\, x = k\, \xi(X)$, which shows K^c-equivariance.

We proceed to show that $\xi^{-1}(M)$ is a realization of G/K as a bounded domain in $\mathcal{C}^+$.

We say that the roots α, β are <u>strongly orthogonal</u>, denoted $\alpha \perp\!\!\!\perp \beta$ if they are not linearly dependent and if $\alpha + \beta$ and $\alpha - \beta$ are not roots.

<u>LEMMA 4.</u> There exists a set Δ of strongly orthogonal roots in Φ such that $\mathcal{O} = \sum_{\alpha \in \Delta} R\, X_\alpha$ is a Cartan subalgebra of the pair $(\mathcal{G}, \mathcal{K})$.

One constructs Δ inductively, always choosing the lowest root in Φ orthogonal to the ones already chosen. We omit the proof which is a direct (though not short) algebraic reasoning to be found in [7] p. 582-583 or [8] p. 314-315 .

LEMMA 5. Let $\mathfrak{l}$ be the Lie algebra over C by H, E^+, E^- and the relations

$$[E^+, E^-] = H, \ [H, E^+] = 2E^+, \ [H, E^-] = -2E^-$$

In any (real) Lie group with Lie algebra $\mathfrak{l}$ we have, for all $t \in R$,

$$\exp t(E^+ + E^-) =$$
$$= \exp(\tanh t) E^+ \cdot \exp(\log\cosh t) H \cdot \exp(\tanh t) E^-$$

Proof. The simply connected group with Lie algebra $\mathfrak{l}$ is $SL(2, \mathbb{C})$, with

$$H = \begin{pmatrix} 1 & 0 \\ 0 & -1 \end{pmatrix} \quad E^+ = \begin{pmatrix} 0 & 1 \\ 0 & 0 \end{pmatrix} \quad E^- = \begin{pmatrix} 0 & 0 \\ 1 & 0 \end{pmatrix}$$

It is enough to check the identity in this group; this is a trivial computation.

LEMMA 6. Let $A = \exp \mathfrak{a}$. Then we have

$$A(x) = \xi\{\sum_{\alpha\in\Delta} t_\alpha E_\alpha \mid |t_\alpha| < 1\}.$$

Proof. Let $g \in A$. Then $g = \exp \sum_{\alpha\in\Delta} t_\alpha X_\alpha$ for some real $\{t_\alpha\}$. Using strong orthogonality of Δ and Lemma 5 we have $g = p^+ k p^-$ with $p^+ = \exp \sum_{\alpha\in\Delta} (\tanh t_\alpha) E_\alpha$ and $kp^- \in K^C P^-$. So $gx = p^+ x$, and the adsertion follows.

THEOREM 4. $M \subset \xi(\mathscr{C}^+)$. $D = \xi^{-1}(M)$ is a bounded domain in $\mathscr{C}^+$. $G_0(D) = \xi^{-1}(G/\mathbb{C})$ (where C denotes the center of G); every element of this group extends to a holomorphic map on a neighborhood of $\overline{D}$. The isotropy group at 0 is $ad(K/C)$.

Proof. We have $G = KAK$, hence $M = G(x) =$ $= KAK(x) = KA(x)$.

By Lemma 6, $\xi^{-1}A(x) = \{\sum_{\alpha\in\Delta} t_\alpha E_\alpha \mid |t_\alpha| < 1\}$.

Since ξ is K-equivariant, $\xi^{-1}(M) = \xi^{-1} K A(x) = K \xi^{-1} A(x) =$
$\left\{ \mathrm{ad}(k) \sum_{\alpha \in \Delta} b_\alpha E_\alpha \,\middle|\, |b_\alpha| < 1,\ k \in K \right\}$.
The other assertions are trivial.

Note . One can study the orbit of $\exp i\mathfrak{a} \subset G_{\mathfrak{u}}$ in a way similar to lemma 6 . With the help of this one can show that $\xi(\mathfrak{p}^+)$ in dense in M^* ([16] p. 286-287).

Some remarks on classification. It is clear now that in order to classify bounded symmetric domains (up to isomorphism) it suffices to classify all pairs $(\mathfrak{g}, \sigma)$ where $\mathfrak{g}$ is compact simple, σ is an involutive automorphism, and $\mathfrak{k}$, the fixed point set of σ , is not semisimple. The classification of all $(\mathfrak{g}, \sigma)$ without any condition on $\mathfrak{k}$ is known (e. g. [8] Ch. IX) and one can simply select out those for which $\mathfrak{k}$ is not semisimple; the problem can be solved, however, with considerably less work. The simplest way is the following, due to J. A. Wolf [23]. Since $\mathfrak{k}$ is one-dimensional and $\mathfrak{f}$ is a Cartan subalgebra of both $\mathfrak{g}$ and $\mathfrak{k}$, it follows that $\mathfrak{g}$ has exactly one simple non compact root: this simple root occurs with coefficient 1 in the roots in Φ , with coefficient 0 in the compact roots. In other words there is a simple root which occurs with coefficient 1 in the maximal root. It is also easy to see that this last condition is sufficient to guarantee a splitting $\mathfrak{g} = \mathfrak{k} + \mathfrak{p}$ under an involution, with $\mathfrak{k}$ not semisimple. So the problem is reduced to the computation of the maximal roots of the simple algebras.

Carrying out the computation (done in [23]) one finds that there are four infinite classes of bounded symmetric domains corresponding to certain classical groups (they are called "classical domains");

and two "exceptional domains" corresponding to exceptional groups.

Explicit descriptions of all these domains are known. The classical domains have been much studied, they are described e.g. in [12] and [18]. The exceptional domain for the group E_7 is discussed in detail in [9], cf. also [3]. The other exceptional domain (corresponding to E_6) is given an explicit realization in [18], § 18, and another one in [10]. A table of the most important constants characterizing all these domains is given in [15].

All this is rather irrelevant from our point of wiew; in these lectures we shall never have to make use of the classification.

3. Boundary structure of D .

The Cayley transform and some lemmas . We use the notation of sec. 2 . We also assume that we are in the irreducible case ; the extensions to the general case are trivial.

For $\alpha \in \Delta$ and for $\Gamma \subset \Delta$ we define

$$c_\alpha = \exp \frac{\pi}{4} iX_\alpha$$

$$c_\Gamma = \prod_{\alpha \in \Gamma} c_\alpha$$

$$c = c_\Delta$$

They are elements of $G^{\mathbb{C}}$, also of G_U.

Lemma 1. For each $\alpha \in \Delta$,

$$\operatorname{ad}(c_\alpha) : \begin{cases} X_\alpha \longmapsto X_\alpha \\ Y_\alpha \longmapsto H_\alpha \\ H_\alpha \longmapsto -Y_\alpha \end{cases}$$

Proof. X_α , Y_α , H_α span a 3-dimensional algebra as in Lemma 2-4; the elemnt corresponding to c_α in SL (2, $\mathbb{C}$) is $\frac{1}{\sqrt{2}} \begin{pmatrix} 1 & i \\ i & 1 \end{pmatrix}$, and acts as stated , by a simple computation.

Define $\mathfrak{f}^-$ as the real space spanned by all iH_α ($\alpha \in \Delta$), and $\mathfrak{f}^+$ as the orthogonal complement of $\mathfrak{f}^-$ in $\mathfrak{f}$ with respect to B .

Lemma 2. ad(c) interchanges $i\mathfrak{f}^-$ and $J\alpha$; it acts trivially on $\mathfrak{f}^+$.

Proof. The first statement follows from Lemma 1 and strong orthogonality of Δ. For the second note that $H \in \mathfrak{h}^+$ implies $\alpha(H) = 0$ $(\alpha \in \Delta)$, since $H_\alpha \in \mathfrak{h}^-$. Hence $[H, X_\alpha] = 0$ for all $\alpha \in \Delta$.

Corollary. $i\,\mathfrak{h}^+ + J\mathfrak{a}$ is a Cartan subalgebra of $\mathfrak{g}$.

ad (c) is an automorphism of $\mathfrak{g}^{\mathbb{C}}$ carrying $\mathfrak{h}^{\mathbb{C}}$ to $(\mathfrak{h}^+) + J\mathfrak{a}^{\mathbb{C}}$, $(\mathfrak{h}^-)^{\mathbb{C}}$ to $J\mathfrak{a}^{\mathbb{C}}$. It follows that the root system of $\mathfrak{g}^{\mathbb{C}}$ with respect to $J\mathfrak{a}$ is the same as the restriction (or projection) of the ordinary $\mathfrak{h}^{\mathbb{C}}$-root system onto $(\mathfrak{h}^-)^{\mathbb{C}}$. The following three lemmas allow a complete description of this root system. (They will be used only at two points, most of what we do is independent of them. Still they clearly contain information of fundamental importance).

Lemma 3. If $\alpha \in \Delta$ and γ is any root, then $\alpha + \gamma$, $\alpha - \gamma$, are not both roots. In particular, $\alpha \perp \gamma$ implies $\alpha \perp\!\!\!\perp \gamma$.

Lemma 4. The restriction to $(\mathfrak{h}^-)^{\mathbb{C}}$ of any $\mathfrak{h}^{\mathbb{C}}$-root is either 0, or $\pm\alpha$, or $\pm\frac{\alpha}{2} \pm \frac{\beta}{2}$, or $\pm\frac{\alpha}{2}$ $(\alpha, \beta \in \Delta, \alpha \neq \beta)$.

The proofs are rather long arguments based on the ordering and on standard properties of roots. They are presented very clearly in [7] on pp. 585-588 which can be read without any preparation.

Lemma 5. The small Weyl group, acting on $i\,\mathfrak{h}^-$ consists of all signed permutations of Δ.

Proof. On the pages referred to above of [7] it is shown that $\frac{\alpha}{2} - \frac{\beta}{2}$ $(\alpha, \beta \in \Delta)$ is an $(\mathcal{J}^-)$-root if and only if $\pm \frac{\alpha}{2} \pm \frac{\beta}{2}$ are (for all combinations of signs). Suppose $\frac{\alpha}{2} - \frac{\beta}{2}$ and $\frac{\beta}{2} - \frac{\gamma}{2}$ are $(\mathcal{J}^-)$-roots. Then their inner product is negative $(-\frac{1}{4} < \beta, \beta >)$, hence $\frac{\alpha}{2} - \frac{\gamma}{2}$ is an $(\mathcal{J}^-)$-root.

It follows that either all combinations $\pm \frac{\alpha}{2} \pm \frac{\beta}{2}$ $(\alpha, \beta \in \Delta)$ are $(\mathcal{J}^-)$-roots, or Δ is the union of subsets Δ', Δ'' such that $\pm \frac{\alpha'}{2} \pm \frac{\alpha''}{2}$ $(\alpha' \in \Delta', \alpha'' \in \Delta'')$ is never a root. The latter case is impossible by irreducibility. (This is the first point where we are really making use of irreducibility).

It follows that the system of restricted roots is one of the classical systems BC_ℓ or C_ℓ (depending on wheter the $\pm \frac{\alpha}{2}$ actually occur). The Weyl group of both of these consists of all signed permutations of Δ.

Boundary components. The map $j = \frac{1}{2}(I - iJ)$ is a (real) linear isomorphism $\mathcal{C} \to \mathcal{C}^+$ such that

$$jX_\alpha = E_\alpha \qquad jY_\alpha = iE_\alpha$$

for $\alpha \in \Phi$ (trivial computation). j commutes with the action of ad K, since J does. Let $\mathfrak{a}^+ = \sum_{\alpha \in \Delta} \mathbb{R} E_\alpha$; it is clear then that $j\mathfrak{a} = \mathfrak{a}^+$.

Under the form B_τ, $\mathcal{G}^{\mathbb{C}}$ has the structure of a complex Euclidean space. The norm $\|T\|$ of any linear transformation T will mean the norm with respect to this structure. (Note that $\{E_\beta\}$ $(\beta \in \Phi)$ is an orthonormal system, and that $ad(G_U)$, in particular ad(c) and ad(K), act by unitary transformations).

Lemma 6. $D = \{ E \in \mathfrak{p}^{+} \mid \| ad(j^{-1}(E)) \| < 2 \}$.

Proof. D is invariant under ad(K), and so is the norm on the right, since $ad(j^{-1}(ad(k)E)) = ad(k)\, ad(j^{-1}(E)) ad(k)^{-1}$ and ad(k) is unitary. Hence it suffices to consider $E \in D \cap \mathfrak{a}^{+}$. By Lemma 2-6 this means $E \in A(0)$, $E = \sum_{\alpha \in \Delta} b_\alpha E_\alpha$ $(|b_\alpha| < 1)$.

Now $j^{-1}(E)$ is transformed by ad(c). J into $\sum b_\alpha H_\alpha$, and $ad(j^{-1}(E))$ has the same norm as $ad(\sum b_\alpha H_\alpha)$, since ad (c) and J are unitary. By Lemma 4 we know that the eigenvalues of the latter are $\pm b_\alpha \pm b_\beta$; and possibly $\pm b_\alpha$ $(\alpha, \beta \in \Delta)$, and it is clear that all eigenvalues are < 2 if and only if $|b_\alpha| < 1$ for all $\alpha \in \Delta$.

We have to make some definitions now. For a subset $\Gamma \subset \Delta$, let $\mathfrak{h}^{-}_{\Gamma}$ denote the real space spanned by iH_α $(\alpha \in \Gamma)$. The centralizer of $\mathfrak{h}^{-}_{\Delta-\Gamma}$ in $\mathfrak{g}^{\mathbb{C}}$ is clearly

$$\mathfrak{h}^{\mathbb{C}} + \sum_{\beta \perp \Delta - \Gamma} \mathbb{C} E_\beta .$$

Denoting by $\mathfrak{g}^{\mathbb{C}}_{\alpha}$ the 3-dimensional algebra generated by H_α, E_α, $E_{-\alpha}$ $(\alpha \in \Delta)$, the centralizer of $\sum_{\Delta-\Gamma} \mathfrak{g}^{\mathbb{C}}_{\alpha}$ is

$$(\mathfrak{h}^{+})^{\mathbb{C}} + (\mathfrak{h}^{-}_{\Gamma})^{\mathbb{C}} + \sum_{\beta \perp\!\!\perp \Delta-\Gamma} \mathbb{C} E_\beta$$

Both centralizers are clearly reductive, and by Lemma 3 have the same derived algebra, which we denote by $\mathfrak{g}^{\mathbb{C}}_{\Gamma}$.

We define $\mathfrak{g}_\Gamma, \mathfrak{g}_{\Gamma,v}, \mathfrak{k}_\Gamma, \mathfrak{b}_\Gamma, \mathfrak{b}^{+}_{\Gamma}, \mathfrak{a}_\Gamma$ as intersections of $\mathfrak{g}^{\mathbb{C}}_{\Gamma}$ with $\mathfrak{g}, \mathfrak{g}_v, \mathfrak{k}$ etc. We denote by G_Γ, $G_{\Gamma,v}$, etc. the corresponding analytic subgroups of $G^{\mathbb{C}}$.

Let $M^*_\Gamma = G^{\mathbb{C}}_\Gamma(x)$, $M = G(x)$ (orbits in M^*).

<u>Lemma 7.</u> M^*_Γ, M_Γ are totally geodesic Hermitian symmetric subspaces of M^*, M respectively; they are isomorphic with $G_{\Gamma, U}/K$ and G_Γ/K_Γ. The following diagram, where j_Γ, ξ_Γ denote the maps of theorems 2-2, 2-3 constructed relative to $G^{\mathbb{C}}_\Gamma$, and the vertical arrows denote inclusions

$$\begin{array}{ccccc} M & \xrightarrow{\ j\ } & M^* & \xleftarrow{\ \xi\ } & \mathfrak{p}^+ \\ \uparrow & & \uparrow & & \uparrow \\ M_\Gamma & \xrightarrow[j_\Gamma]{} & M^*_\Gamma & \xleftarrow[\xi_\Gamma]{} & \mathfrak{p}^+_\Gamma \end{array}$$

is commutative. In particular, $D_\Gamma = D \cap \mathfrak{p}^+_\Gamma$ is the standard realization of G_Γ/K_Γ as a bounded domain.

Proof. The algebras $\mathfrak{g}_\Gamma$, $\mathfrak{g}_{\Gamma,U}$ are invariant under σ and J by construction; the same is true after exponentiation in $G^{\mathbb{C}}$. This proves the first statement; the rest is tedious but trivial checking.

As an abbreviation, we write E_Γ for $\sum_{\alpha\in\Gamma} E_\alpha$. Also, for $E \in \mathfrak{p}^+$, we write $c_\Gamma \cdot E$ instead of $\xi^{-1} c_\Gamma \xi(E)$. (This has nothing to do with $\mathrm{ad}(c_\Gamma)E$.)

<u>Lemma 8.</u> For $E \in D_\Gamma$, $c_{\Delta-\Gamma}\cdot E = iE_{\Delta-\Gamma} + E$. In particular, $c_{\Delta-\Gamma}\cdot 0 = iE_{\Delta-\Gamma}$.

Proof. $c_{\Delta-\Gamma}$ is in the exponential of $\sum_{\Delta-\Gamma} \mathfrak{g}^{\mathbb{C}}_\alpha$, so it commutes with G_Γ. This reduces the first assertion to the second. The second is reduced by strong orthogonality to the obvious identity in $SL(2, \mathbb{C})$:

$$\frac{1}{\sqrt{2}}\begin{pmatrix}1 & i\\ i & 1\end{pmatrix} = \begin{pmatrix}1 & i\\ 0 & 1\end{pmatrix}\begin{pmatrix}\sqrt{2} & 0\\ 0 & \frac{1}{\sqrt{2}}\end{pmatrix}\begin{pmatrix}1 & 0\\ i & 1\end{pmatrix}$$

(whence $c_{\Delta-\Gamma}(x) = \xi(iE_{\Delta-\Gamma})$, etc.).

It is now clear that $c_{\Delta-\Gamma}\cdot D_\Gamma$ is on the boundary of D, and that the intersection of the affine subspace $iE_{\Delta-\Gamma}+\mathcal{C}_\Gamma^+$ with $\overline{D}$ is $c_{\Delta-\Gamma}\cdot\overline{D}_\Gamma$. However, we want the following stronger result.

<u>Lemma</u> 9. $c_{\Delta-\Gamma}\cdot\overline{D}_\Gamma$ is the intersection of $\overline{D}$ with the (real or complex) affine hyperplane passing through and orthogonal to $iE_{\Delta-\Gamma}$ in $\mathcal{C}^+$.

Proof. By Lemma 6 the statement amounts to showing that if $V = Y_{\Delta-\Gamma} + U \in \mathcal{C}$, $Y_{\Delta-\Gamma} \perp U$, and $\|\operatorname{ad} V\| \leq 2$, then necessarily $U \in \mathcal{C}_\Gamma$.

We can write $U = U_1 + U_2$ where U_1 is orthogonal to each Y_α $(\alpha \in \Delta-\Gamma)$, and $U_2 = \sum_{\Delta-\Gamma} a_\alpha Y_\alpha$. Let $W_\alpha = X_\alpha + iH_\alpha$ and $W = \sum_{\Delta-\Gamma} W_\alpha$. One computes then (using Lemma 1) that $[Y_\alpha, W] = 2W_\alpha$, $\operatorname{ad}(V)W = 2W + [U_1, W]$ (since $[U_2, W] = 0$, from $U_2 \perp Y_{\Delta-\Gamma}$) and $[U_1, W] \perp W$.

$\|\operatorname{ad} V\| \leq 2$ now implies $[U_1, W] = 0$. But $[U_1, W] = i\,[U_1, H_{\Delta-\Gamma}] + T$, with $T \in \mathcal{K}$, and hence $[U_1, H_{\Delta-\Gamma}] = 0$. So U_1 is in the centralizer of $H_{\Delta-\Gamma}$.

Application of Lemma 4 now shows that U_1 must be in the centralizer of each H_α $(\alpha \in \Delta - \Gamma)$, finishing the proof.

A set of the form $\mathrm{ad}(k)\, c_{\Delta-\Gamma} \cdot D_\Gamma$ $(k \in K, \Gamma \subsetneqq \Delta)$ is called a <u>boundary component</u> of D. Obviously it is contained in the boundary ∂D of D.

<u>Theorem</u> 1, (i) ∂D is the union of mutually disjoint boundary components.

(ii) $c_{\Delta-\Gamma'} \cdot D_{\Gamma'} = \mathrm{ad}(k) c_{\Delta-\Gamma} \cdot D_\Gamma$ if and only if Γ and Γ' have the same number of elements.

(iii) If $f : U \longrightarrow \overline{D}$ is a holomorphic curve (here U is the complex unit disc) and F is a boundary component then $f(U) \cap F = \emptyset$ implies $f(U) \subset F$. Each F is minimal with respect to this property.

(iv) The orbits of G on the boundary are $G(iE_{\Delta-\Gamma}) = \bigcup_{k \in K} \mathrm{ad}(k) c_{\Delta-\Gamma} D_\Gamma$. There are ℓ such orbits (ℓ = rank of D).

<u>Proof</u>. (i) Since $D = KA(0)$, any point E in $\overline{D}$ can be transformed by some $k \in K$ into $\overline{A(0)}$ and also into $i\overline{A(0)}$; if $E \in \partial D$, $\mathrm{ad}(k)E$ is on $\partial iA(0)$, so of the form $iE_{\Delta-\Gamma} + \sum_\Gamma b_\alpha E_\alpha$ $(|b_\alpha| < 1)$. This shows that $E \in \mathrm{ad}(k^{-1}) c_{\Delta-\Gamma} \cdot D_\Gamma$.

(ii) For this we have to see only that $E_{\Delta-\Gamma'} = \mathrm{ad}(k) E_{\Delta-\Gamma}$ (some $k \in K$) if and only if Γ, Γ' have the same number of elements. Using the K-equivariant map j this reduces exactly to Lemma 5.

(iii) We may assume $F = c_{\Delta - \Gamma} \cdot D_{\Gamma}$. Suppose $\phi(U)$ intersects it. Let λ be the complex coordinate function in the direction $iE_{\Delta-\Gamma}$, $|\lambda \circ \phi|$ assumes then its maximum, and is therefore constant. It follows that $\phi(U)$ is contained in the hyperplane through $iE_{\Delta-\Gamma}$, and, by Lemma 9, in $c_{\Delta-\Gamma} \overline{D}_{\Gamma}$. Now $\phi(U)$ cannot intersect $\partial c_{\Delta-\Gamma} D_{\Gamma}$, because then it would have to stay entirely in a boundary component of $c_{\Delta-\Gamma} D_{\Gamma}$. The statement on minimality is clear, for any two points of F can be joined by a holomorphic curve.

(iv) The orbit $iE_{\Delta-\Gamma}$ under $\exp \mathfrak{a}_{\Gamma}$ is $\left\{ iE_{\Delta-\Gamma} + \sum_{\Gamma} b_{\alpha} E_{\alpha} \,\middle|\, |b_{\alpha}| < 1 \right\}$ as one sees at once from Lemma 2-6 and Lemma 7. (In fact one can show easily that this is also the orbit under A.) It follows that $G(iE_{\Delta-\Gamma})$ contains $ad(K)c_{\Delta-\Gamma} \cdot D_{\Gamma}$. To see that it contains nothing else, note that by the caracterization given in (iii) every automorphism maps boundary components isomorphically onto boundary components. The rank of D is equal to the number of elements in Γ, and must be preserved.

<u>Corollary</u>. The union of all 0-dimensional boundary components is $S = G(iE_{\Delta}) = K(iE_{\Delta})$; it is the only orbit of G which is also an orbit of K, and it is equal to the Bergman-Shilov boundary of D.

The last statement means that every continuous function $f : \overline{D} \rightarrow \mathbb{C}$ which is holomorphic in D assumes its maximum (in modulus) on S; this is easy to see by approximating the function $f(z)$ by functions $f(rz)$ $(0 < r < 1)$, and using the maximum principle on the boundary components. It can also be seen without knowing about boundary components, as it will turn out later.

S as a homogeneous space. This section does not make use the machinery of Lemmas 3 to 9. We know that $G(iE_\Delta) = K(iE_\Delta)$, this set is called S, we now want to study the isotropy groups F and L of G and K, respectively, at iE_Δ. F, and even F_0 are transitive on D, since $G/F = K/L$ implies that $G = FK = F_0K$.

We define $Z_1 = \frac{i}{2} H_\Delta$ (as usual, H_Δ stands for $\sum H_\alpha$, $\alpha \in \Delta$) and $Z_2 = Z - Z_1$.

Z_1, iX, iY span a 3-dimensional simple real Lie algebra $\mathfrak{m} \subset \mathfrak{g}_v \subset \mathfrak{g}^{\mathbb{C}}$, $c \in \exp \mathfrak{m}$, exp referring to $\mathfrak{g}^{\mathbb{C}}$.

$\mathrm{ad}_{\mathfrak{g}^{\mathbb{C}}}(\exp \mathfrak{m})$ can be regarded as a representation of the simply connected group SU(2). The element of SU(2) corresponding to c is $\frac{1}{\sqrt{2}}\begin{pmatrix} 1 & i \\ i & 1 \end{pmatrix}$, it has order 8.

Every irreducible representation of SU(2) has kernel $\{I\}$ or $\{I, -I\}$; accordingly the image of c has order 8 or 4.

Note that $\sigma\, \mathrm{ad}(c)\, \sigma = \sigma\, \mathrm{ad}(c)\, \sigma^{-1} = \mathrm{ad}(\sigma^* c) = $
$= \mathrm{ad}(\exp i \frac{\pi}{4} \sigma(X)) = \mathrm{ad}(\exp(-i \frac{\pi}{4} X)) = \mathrm{ad}(c)^{-1}$. Hence $\mathrm{ad}(c)^4$ commutes with σ, so it preserves $\mathfrak{p}^{\mathbb{C}}$, $\mathfrak{k}^{\mathbb{C}}$. It also preserves $\mathfrak{p}$, $\mathfrak{k}$ since it preserves $\mathfrak{g}_v$. We define

$\mathfrak{p}_1, \mathfrak{p}_2$ the (± 1)-eigenspaces of $\mathrm{ad}(c)^4$ in $\mathfrak{p}$

$\mathfrak{k}', \mathfrak{g}_2$ " in $\mathfrak{k}$

$\mathfrak{k}_1 = [\mathfrak{p}_1, \mathfrak{p}_1]$

$\mathfrak{g}_1 = \mathfrak{k}_1 + \mathfrak{p}_1$

$\mathfrak{l}_2$ the centralizer of $\mathfrak{g}_1$ in $\mathfrak{k}' + \mathfrak{p}_1$

$\mathfrak{p}_1^{\mathbb{C}}$, $\mathfrak{p}_1^+$, etc., in the obvious way.

$\mathfrak{k}' + \mathfrak{b}_1$ is σ-invariant; it is a sub-oiLa of $(\mathfrak{g}, \sigma, Q)$ with the exception that it need not satisfy (iii) of the definition in sec. 1. This, however is irrelevant for the decomposition, all it changes is that to the sum $\mathfrak{g}_0 \oplus \cdots \oplus \mathfrak{g}_t$ there is added another summand, on which σ acts trivially. It follows that $\mathfrak{k}' + \mathfrak{b}_1 = \mathfrak{l}_2 \oplus \mathfrak{g}_1$, and $\mathfrak{k}' = \mathfrak{l}_2 + \mathfrak{k}_1$.

<u>Lemma</u> 10. On $\mathfrak{b}_1$, $ad(Z_1) = ad(Z)$. On $\mathfrak{b}_2$, $ad(Z_1) = \frac{1}{2} ad(Z)$.

<u>Proof.</u> $2Z_1$ is the image of iX under an automorphism of $\mathfrak{m}$. By the remark made on the kernels of the representations of $SU(2)$, $\exp(\frac{\pi}{2} Z_1)$ and $\exp(\frac{\pi}{4} iX) = c$ have the same order on any irreducible representation space in $ad_{\mathfrak{g}^{\mathbb{C}}}(\exp \mathfrak{m})$. The sum of the irreducible spaces where they have order 8 is $\mathfrak{b}_2 + \mathfrak{g}_2^{\mathbb{C}}$ by definition. $Z_1 \in \mathfrak{f}$ is represented by diagonal matrices, so on $\mathfrak{g}_2^{\mathbb{C}} + \mathfrak{b}_2^{\mathbb{C}}$ we have $ad \exp(\frac{\pi}{2} Z_1) = e^{\pm\frac{\pi}{4} i} I$. Hence $ad(Z_1) = \pm \frac{i}{2}$ there, and using the ordering of roots it follows that $ad(Z_1) = \pm \frac{i}{2}$ on $\mathfrak{g}_2^{\pm} + \mathfrak{b}_2^{\pm}$. This implies, in particular, $ad(Z_1) = \frac{1}{2} ad(Z)$ on $\mathfrak{b}_2^{\mathbb{C}}$. The same argument with the irreducible spaces where $ad(c)$ has order 4 gives the statement about $\mathfrak{b}_1$.

We denote by G_1, K_1 the analytic subgroups of $G^{\mathbb{C}}$ corresponding to $\mathfrak{g}_1$, $\mathfrak{k}_1$.

<u>Lemma</u> 11. $G_1(x) = G_1/K_1$ is a totally geodesic Hermitian symmetric subspace of M^*. Its standard realization as a bounded domain is $D \cap \mathfrak{b}_1^+$.

Proof. By the preceding $\mathfrak{g}_1$ splits under σ as $\mathfrak{k}_1 + \mathfrak{p}_1$, and is invariant under $J = \mathrm{ad}(Z)$. The rest is simple direct checking.

Note that it may happen that $\mathfrak{g}_1 = \mathfrak{g}$. We say then that G/K is of tube type (it will be seen later that it is in this case that the Cayley transform of D is a tube domain).

$\mathfrak{k}_1$, $\mathfrak{p}_1$ are preserved by $\mathrm{ad}(c)^2$, by an argument seen before (note that on $\mathfrak{g}_1$ $\mathrm{ad}(c)^4 = I$). We define:

$$\mathfrak{l}_1, \mathfrak{l}_2 \text{ the } (\pm 1)\text{-eigenspaces of } (\mathrm{ad}(c))^2 \text{ in } \mathfrak{k}$$
$$\mathfrak{l} = \mathfrak{l}_1 + \mathfrak{l}_2 .$$

Lemma 12. The Lie algebra of L is $\mathfrak{l}$. The Lie algebra of L_1, the isotropy group of K_1 at iE, is $\mathfrak{l}_1$.

Proof. Since $\xi(iE_\Delta) = cx$, the Lie algebra of L is $\mathfrak{k} \cap \mathrm{ad}(c)(\mathfrak{k}^C + \mathfrak{p}^-)$. Denote it by $\mathfrak{l}'$. Since $\mathfrak{k} \subset \mathfrak{g}_U$, since ad(c) preserves $\mathfrak{g}_U$, and since $(\mathfrak{k}^C + \mathfrak{p}^-) \cap \mathfrak{g}_U = \mathfrak{k}$, we have $\mathfrak{l}' = \mathfrak{k} \cap \mathrm{ad}(c)\,\mathfrak{k}$. This means those $V \in \mathfrak{k}$ for which $\mathrm{ad}(c)\,\sigma\,\mathrm{ad}(c)^{-1} V = V$.

Using $\sigma\,\mathrm{ad}(c)\,\sigma = \mathrm{ad}(c)^{-1}$ and $V = \sigma V$, this condition means $\mathrm{ad}(c)^2 V = V$, i.e. $V \in \mathfrak{l}$. For L_1 the proof is similar.

Remarks. $\mathrm{ad}(c)^2$ is an involutive automorphism of $\mathfrak{k}_1$; it is easy to see that there is a corresponding (real) symmetric space $K_1(iE_\Delta) = K_1/L_1$, totally geodesic in M^*. This means that if G/K is of tube type then S is a compact) symmetric space. In the general case one shows easily ([15] Th. 4.9) that $K(c^2x) = K/_{K'}$ is a Hermitian symmetric subspace of M^*, and S is a fiber space over it with fiber K_1/L_1.

Lemma 13. (i) $ad(c)^2$ interchanges $i\mathcal{C}_2$ with $\mathcal{Y}_2$, $\mathcal{C}_2^{\pm}$ with $\mathcal{Y}_2^{\pm}$.

(ii) $ad(c)$ interchanges the (-1)-eigenspace of $ad(c)^2$ in $i\mathcal{C}_1$ with $\mathcal{Y}_1$.

(iii) J interchanges the (± 1)-eigenspaces of $ad(c)^2$ in $i\mathcal{C}_1$.

Proof. (i) Let $V \in i\mathcal{C}_2$. Then $\sigma(ad(c)^2V) = ad(c)^{-2}\sigma V = -ad(c)^{-2}V = -(ad\, c)^6V = ad(c)^2V$, hence $ad(c)^2$ carries $i\mathcal{C}_2$ into $\mathcal{Y}_2$. The proof of the converse is similar. The second half of (i) amounts to a convection: We may order the roots so that $-i\alpha(Z_1) > 0$ implies $\alpha > 0$ (this is in accordance with the former convention that $-i\alpha(Z) > 0$ implies $\alpha > 0$).

(ii) is proved in the same way as (i). To show (iii) note that if $ad(c)^2 V = \pm V$, then $ad(c)^2 JV = -J\, ad(c)^2V = \mp JV$ (where we used $ad(c)^2 Z_1 = -Z_1$, which is easy to check from the definitions).

Now the last definitions:

$$\mathcal{U}_1^{\pm} = \mathcal{U}_1^{\pm} \cap ad(c)$$

$$\mathfrak{u}_2^{\pm} = (\mathfrak{c}_2^{\pm} + \mathfrak{q}_2^{\pm}) \cap \mathrm{ad}(c)\mathfrak{g}$$

$$\mathfrak{u}^{\pm} = \mathfrak{u}_1^{\pm} + \mathfrak{u}_2^{\pm}$$

$$\mathfrak{k}'^{*} = \mathfrak{l} + i\,\mathfrak{q}_1$$

$$\mathfrak{k}_1^{*} = \mathfrak{l}_1 + i\,\mathfrak{q}_1$$

Lemma 14. (i) ad(Y) is semisimple on with eigenvalues 0, ± 1, ± 2 and corresponding eigenspaces $\mathrm{ad}(c)^{-1}\mathfrak{k}'$, $\mathrm{ad}(c)^{-1}\mathfrak{u}_2^{\pm}$, $\mathrm{ad}(c)^{-1}\mathfrak{u}_1^{\pm}$.

(ii) $\mathfrak{u}_1^{\pm}$ are real forms of $\mathfrak{c}_1^{\pm}$

$\mathfrak{u}_2^{\pm}$ " " " " $\mathfrak{c}_2^{\pm} + \mathfrak{q}_2^{\pm}$

(iii) $[\mathfrak{u}_2^{+}, \mathfrak{u}_2^{+}] \subset \mathfrak{u}_1^{+}$, $[\mathfrak{u}_1^{+}, \mathfrak{u}_1^{+}] =$

$= [\mathfrak{u}_1^{+}, \mathfrak{u}_2^{+}] = 0$, so $\mathfrak{u}^{+}$ is step 2 nilpotent.

Proof. ad(Y) is semisimple on $\mathfrak{g}^C$ and preserves $\mathfrak{g}$. Hence its eigenspaces on are real forms of its eigenspaces on $\mathfrak{g}^C$. We know that $\mathrm{ad}(c)^{-1} Z_1 = \frac{i}{2} Y$, and that the eigenvalues of $\mathrm{ad}(Z_1)$ are 0, $\pm \frac{i}{2}$, $\pm i$ with eigenspaces $\mathfrak{k}'^C$, $\mathfrak{c}_2^{\pm} + \mathfrak{q}_2^{\pm}$, $\mathfrak{c}_1^{\pm}$. Hence ad(Y) has eigenvalues 0, ± 1, ± 2 with eigenspaces $\mathrm{ad}(c)^{-1}\mathfrak{k}'^C$, etc.

Remains to check that $\mathrm{ad}(c)^{-1}\mathfrak{k}'^C \cap \mathfrak{g} = \mathfrak{k}'^{*}$, but that is clear from Lemma 13 (ii). We also see that $\mathrm{ad}(c)^{-1}\mathfrak{u}_1^{\pm}$

are real forms of $\mathrm{ad}(c)^{-1}\,\mathfrak{u}_1^{\pm}$, whence (ii) follows. (iii) is obvious by the general rules about brackets of eigenspaces (of ad(Y) in this case).

Theorem 2. (i) The Lie algebra of F is $\mathfrak{f} = \mathrm{ad}(c)^{-1}\mathfrak{k}'^{*} + \mathrm{ad}(c)^{-1}\mathfrak{n}^{+}$. It is the normalizer of $\mathrm{ad}(c)^{-1}\mathfrak{n}^{+}$ in $\mathfrak{g}$.

(ii) K'^{*}, N^{+}, the analytic subgroups for $\mathfrak{k}'$, $\mathfrak{n}^{+}$ are closed in G^{C}. $F_o = \mathrm{ad}(c)^{-1}K'^{*} \cdot \mathrm{ad}(c)^{-1}N^{+}$ is a semi-direct product.

Proof. The Lie algebra of F is $\mathfrak{f} = \mathrm{ad}(c)(\mathfrak{k}^{C} + \mathfrak{c}^{-}) \cap \mathfrak{g}$. $\mathrm{ad}(c)^{-1}\,\mathfrak{k}'^{*}$ and $\mathrm{ad}(c)^{-1}\mathfrak{n}^{+}$ are in $\mathfrak{g}$; if we can show that they are also in $\mathrm{ad}(c)(\mathfrak{k}^{C} + \mathfrak{c}^{-})$, then the first statement follows by a dimension count. Now $\mathrm{ad}(c)^{-1}\,\mathfrak{k}'^{*} = \mathrm{ad}(c)\,\mathfrak{k}'^{*} \subset \mathrm{ad}(c)\,\mathfrak{k}^{C}$. $\mathrm{ad}(c)^{-1}\mathfrak{n}_1^{+} \subset \mathrm{ad}(c)^{-1}\,\mathfrak{c}_1^{+} = \mathrm{ad}(c)\,\mathfrak{c}_1^{-}$ since $\mathrm{ad}(c)^2\,\mathfrak{c}_1^{-} = \mathfrak{c}_1^{+}$.

Finally $\mathrm{ad}(c)^{-1}\,\mathfrak{n}_2^{+} \subset \mathrm{ad}(c)^{-1}(\mathfrak{c}_2^{+} + \mathfrak{g}_2^{+}) \subset \mathrm{ad}(c)(\mathfrak{c}_2^{-} + \mathfrak{g}_2^{-})$ by Lemma 13 (i).

The statement about the normalizer follows from Lemma 14 (i). (ii) is proved by a standard kind of argument, the same as Lemma 2-3.

A. Koranyi

4. Generalized halfplanes.

This section is entirely independent of the three preceding ones and has an elementary character.

Let W be a finite-dimensional real vector space. A subset $\Omega \subset W$ is called a regular cone if it is non-empty, open, , convex and such that $0 \neq y \in \bar{\Omega}$, $\lambda > 0$ imply $\lambda y \in \bar{\Omega}$ and $-y \notin \bar{\Omega}$. The dual cone is defined by $\Omega' = \{\alpha \in W' \mid \langle \alpha, y \rangle > 0, \forall 0 \neq y \in \bar{\Omega}\}$ It is not hard to show that Ω' is also a regular cone, and $\Omega'' = \Omega$.

Let now V_1, V_2 be complex vector spaces of dimension $n_1 \neq 0$, n_2 respectively; let W be a real form of V_1, and let be a regular cone in W. An Ω - Hermitian form is a function $\Phi : V_2 \times V_2 \longrightarrow V_1$ such that

(i) Φ is complex linear in the first argument,

(ii) $\Phi(v, u) = \overline{\Phi(u, v)}$ (conjugation with respect to W),

(iii) $\Phi(u, u) \in \bar{\Omega}$ for all $u \in V_2$,

(iv) $\Phi(u, u) = 0$ implies $u = 0$.

The generalized halfplane corresponding to Ω and is then defined by

$$D = D(\Omega, \Phi) = \{(z_1, z_2) \in V_1 \times V_2 \mid \operatorname{Im} z_1 - \Phi(z_2, z_2) \in \Omega\}$$

(This notion is due to Pyatetskii-Shapiro; he calls D a Siegel domain of type I or II depending on whether $n_2 = 0$ or not. If $n_2 = 0$, there is no Φ, and D is simply the 'tube domain' over Ω.)

The distinguished boundary of D is

$$B = \{ (z_1, z_2) \in V_1 \times V_2 \mid \operatorname{Im} z_1 - \Phi(z_2, z_2) = 0 \}.$$

It is not hard to show directly that if f is bounded continuous on $\overline{D}$ and holomorphic on D, then $\sup_D |f| = \sup_B |f|$, and B is a minimal subset of $\overline{D}$ with this property. This will, however be an immediate consequence of some later results.

D admits a group $\mathcal{N}$ of affine holomorphic automorphism, whose elements are pairs $g = (a, c) \in W \times V_2$, acting by

$$g : \begin{cases} z_1 \longrightarrow z_1 + a + 2i\,\Phi(z_2, c) + i\,\Phi(c, c) \\ z_2 \longrightarrow z_2 + c \end{cases}$$

It is easy to check that $\mathcal{N}$ is simply transitive on B, and nilpotent of step 2 (if $n_2 = 0$ it is even Abelian).

It is of interest to consider the group $G(\Omega) =$
$= \{ g_1 \in GL(W) \mid g_1 \Omega = \Omega \}$ and the group $G(\Omega, \Phi) =$
$= \{ (g_1, g_2) \in G(\Omega) \times GL(V_2) \mid g_1 \Phi(u, v) = \Phi(g_2 u, g_2 v), \ \forall u, v \}$.

D is called affine-homogeneous if $pr_1 G(\Omega, \Phi)$ (the projection onto $G(\Omega)$) is transitive on Ω; in this case the group generated by $\mathcal{N}$ and $G(\Omega, \Phi)$ is transitive on D.

All we do in this section is valid for any generalized halfplane. Let us note that $G(\Omega, \Phi)$ always contains at least the one-parameter group $\{ (e^{2t} I, e^{t} I) \mid t \in \mathbb{R} \}$.

In the following we assume that $V_1 = \mathbb{C}^{n_1}$, $V_2 = \mathbb{C}^{n_2}$ with their standard complex Euclidean structure. This is actually quite unimportant, all it is good for is to fix the normalization of certain Haar measures.

We define the measure β on B by $\int_B f(u)\, d\beta(u) =$

$$= \int_{W \times V_2} f(x_1 + i\, \Phi(z_2, z_2), z_2)\, dx_1\, dx_2\, dy_2$$

$g \longrightarrow g \cdot 0$ is a diffeomorphism $\mathcal{N} \to B$, equivariant for the left action of $\mathcal{N}$.

Lifting β to $\mathcal{N}$ under this map, it is immediate to check that we get a Haar measure on $\mathcal{N}$ (both left and right invariant, of course, since $\mathcal{N}$ is nilpotent).

For $0 < p \leq \infty$ we denote by $L^p(B)$ the L^p - space with respect to β.

For a function $f : D \longrightarrow \mathbb{C}$ and $t \in \Omega$ we define $f_t : D \longrightarrow \mathbb{C}$ by

$$f_t(z_1, z_2) = f(z_1 + it, z_2).$$

For $0 < p < \infty$ we define $H^p(D)$ as the space of holomorphic functions of f on D such that

$$\| f \|_p = \sup_{t \in \Omega} \| f_t \,|\, B \|_{L^p(B)} < \infty .$$

We define $H^{\infty}(D)$ as the space of bounded holomorphic functions on D.

The following is a variant of a theorem of Gindikin [6] and plays a fundamental role in this section. For fixed $\lambda \in \Omega'$ we denote $B_\lambda(u, v) = 4 < \lambda, \Phi(u, v) >$; B_λ is an ordinary Hermitian quadratic form on V_2.

<u>Theorem 1</u>. For $z = (z_1, z_2) \in D$, $\lambda \in \Omega'$, $\alpha \in R^{n_2}$, let

$$\overline{\chi_z(\lambda,\alpha)} = (\det B_\lambda)^{3/4}\, e^{2\pi i\langle\lambda,\, z_1+2\sqrt{2}\,\Phi(z_2,\alpha)-i\Phi(z_2,\bar{z}_2)+i\Phi(\alpha,\alpha)\rangle} .$$

Then the map

$$\varphi \longrightarrow f(z) = \int_{\Omega'\times R^{n_2}} \varphi(\lambda,\alpha)\,\chi_z(\lambda,\alpha)\,d\lambda\, d\alpha$$

is a Banach space isomorphism $L^2(\Omega'\times R^{n_2}) \longrightarrow H^2(D)$. (In particular, $H^2(D)$ is a Hilbert space.)

The proof is too long and technical to be given here. We should note, however, that if D is the upper halfplane in $\mathbb{C}$, then this is a classical result of Paley and Wiener. In the tube case $(n_2 = 0)$ it was proved by Bochner [2] . The general case in a somewhat different form and with only a sketchy proof is in [6] .

The present version in due to E. M. Stein and the author (to be published) .

A. Koranyi

Theorem 2. If $f \in H^2(D)$, then $\lim_{\substack{t \to 0 \\ t \in \Omega}} f_t \big|_B$ exists in $L^2(B)$.

Sketch of proof : Let f_o be the function defined on B by setting $y_1 = \Phi(z_2, z_2)$ in the formula of Theorem 1. By using the Plancherel theorem twice, one sees that

$$\|f_t - f_o\|_{L^2(B)} = \int_{\Omega' \times R^{n_2}} |\varphi(\lambda, \alpha)|^2 (1 - e^{-4\pi\langle\lambda, y_1\rangle}) d\alpha \, d\lambda .$$

From here the statement follows by the Lebesgue dominated convergence theorem.

Remark . This theorem gives a manageable way to compute inner products of functions in $H^2(D)$; one only has to compute the inner product in $L^2(B)$ of their boundary functions .

The following short discussion is of a quite general nature; it is probably rather familiar to the reader who has studied the question of the existence of the Bergman kernel and Bergman metric.

Let H be a Hilbert space of complex-valued functions defined on a set E. $S : E \times E \longrightarrow \mathbb{C}$ is called a reproducing kernel for H if, defining $S_w : E \longrightarrow \mathbb{C}$ by $S_w(z) = S(z, w)$ ($w \in E$ fixed), we have $(f, S_w) = f(w)$ for all $f \in H$, $w \in E$.

It is then automatic that $\overline{S(z,w)} = S(w,z)$, and $S(z,z) \geqq 0$ with equality only if $f(z) = 0$ for all $f \in H$.

Suppose now that $\mathcal{H}$ is another Hilbert space, and suppose it contains a family of elements χ_z ($z \in E$) such that $\varphi \mapsto \hat{\varphi}$, where $\hat{\varphi} : E \longrightarrow \mathbb{C}$ is defined by $\hat{\varphi}(z) = (\varphi, \chi_z)$, is an isomorphism $\mathcal{H} \to H$. We claim that S defined by $S(z,w) = (\chi_w, \chi_z)$ is then a reproducing kernel for H .

In fact, defining S_w as above, we have $S_w(z) = S(z,w) = (\chi_w, \chi_z) = \hat{\chi}_w(z)$. Any $f \in H$ equals $\hat{\varphi}$ for some $\varphi \in \mathcal{H}$, and we have $(f, S_w) = (\hat{\varphi}, \hat{\chi}_w) = \hat{\varphi}(w) = f(w)$, proving the claim.

All this applies, of course, to the situation of theorem 1, and after an easy computation we have :

Theorem 3 . $H^2(D)$ has a reproducing kernel, called the Szegö kernel, given by

$$S_w(z) = S(z,w) = \int_{\Omega'} e^{-2\pi\langle\lambda, \rho(z,w)\rangle} (\det B_\lambda) d\lambda ,$$

where $\rho(z,w) = -i(z_1 - \bar{w}_1) - 2\Phi(z_2, w_2)$.

Corollary . For fixed $w \in D$, S_w extends to $B \cup D$ as a bounded continuous function (in fact, S_w is even analytic on $\bar{D}$).

Now we define the Poisson kernel of D by

$$P(u,z) = \frac{|S(u,z)|^2}{S(z,z)} \qquad (u \in B, \quad z \in D) .$$

We shall also use the notation $P_z(u) = P(u,z)$, so that, for fixed $z \in D$, P_z ia a function on B .

Lemma 1 . Let F be a bounded holomorphic function on $\bar{D}$. Then, for all $z \in D$,

$F(z) = (F|_B , P_z)$.

Proof. Since F is bounded and $S_z \in H^2(D)$, we have $FS_z \in H^2(D)$. FS_z has continuous boundary values on B. By Theorem 3,

$$(F|_B, P_z) = \frac{1}{S(z, z)} (F|_B, \bar{S}_z S_z|_B) =$$

$$= \frac{1}{S(z, z)} (FS_z|_B, S_z|_B) =$$

$$= \frac{1}{S(z,z)} (FS_z)(z) = F(z) .$$

Theorem 4. (a) $P(u, z) \geqq 0$ for all $u \in B$, $z \in D$.

(b) For all fixed $z \in D$, P_z is a continuous function on B, vanishing at infinity.

(c) $\int_B P_z(u)\, d\beta(u) = 1$ for all $z \in D$.

(d) For every $u_o \in B$ and every neighborhood N of u_o in B,

$$\lim_{z \to u_o} \int_{u \notin N} P_z(u)\, d\beta(u) = 0 .$$

(e) Denoting, for real s and $z \in \bar{D}$, $z^s = (sz_1, \sqrt{s}\, z_2)$, we have, for all $u \in B$, $z \in D$,

$$s^n P(z^s, u^s) = P(z, u)$$

where $n = n_1 + n_2$,

Proof. (a) is clear from the definition. The first

part of (b) follows from the corollary of Theorem 3, the second part can be deduced from a similar property of S_z (which, in turn, is a consequence of the simple fact that $S_w \in H^2(D)$ for all w we omit the proof). (c) follows by applying Lemma 1 to the function $F \equiv 1$.

To prove (d) we take a function F, holomorphic on $\overline{D}$, such that $|F| \leq 1$ on $\overline{D}$, $F(u_o) = 1$ and $m = \sup_{u \notin N} |F(u)| < 1$. (It is easy to check that $F = cS_{n_o + iy}$, for any $y = (y_1, 0)$, $y_1 \in \Omega$ and appropriately chosen c has these properties.) By Lemma 1,

$$\lim_{z \to u_o} (F, P_z) = \lim_{z \to u_o} F(z) = 1 .$$

On the other hand,

$$|(F, P_z)| = \left| \int_{u \in N} F(u) P(u, z) d\beta(u) + \int_{u \notin N} F(u) P(u, z) d\beta(u) \right| \leq$$

$$\leq \int_{u \in N} P(u, z) d\beta(u) + m \int_{u \notin N} P(u, z) d\beta(u).$$

Using (c) it follows that

$$\lim_{z \to u_o} (1-m) \int_{u \in N} P(u, z) d\beta(u) = 0$$

which proves (d).

The proof of (e) is a straighforward checking, based on the formula of Theorem 3.

<u>Remark</u>. For $t \in \Omega$ we define $\tilde{P}^t : \mathcal{N} \to R$ by

$$\tilde{P}^t(g) = P(0, g \cdot (it, 0)),$$

and for $f : B \to \mathbb{C}$ we define $\tilde{f} : \mathcal{N} \to \mathbb{C}$ by $\tilde{f}(g) = f(g \cdot 0)$. Then it is easy to check that, denoting by F the Poisson integral of f $(F_t(z) = (f, P_z))$, we have

$$\tilde{F}_t(g) = F_t(g \cdot 0) = (\tilde{f} * \tilde{P}^t)(g)$$

where $*$ denotes convolution on $\mathcal{N}$.

Theorem 5. Let f be a function or a measure on B, and let F be its Poisson integral. Then $f = \lim_{t \to 0} F_t$

(a) pointwise if f is continuous and bounded,

(b) in L^p, if $f \in L^p(B)$ $(1 \leq p < \infty)$

(c) in the weak-$*$ topology of L^∞ if $f \in L^\infty(B)$

(d) in the weak-$*$ topology of measures if f is a finite Baire measure.

The proof is immediate from Theorem 4 by standard methods of functional analysis.

The following generalization of Fatou's thorem is somewhat more difficult.

For $g = (a, c) \in \mathcal{N}$ we define $|g| = \mathrm{Max}\{|a|, |c|^2\}$. For $\alpha > 0$ and a cone ω such that $\bar{\omega} \in \Omega$ we define the "admissible domains"

$$\Gamma_{\alpha,\omega}(0) = \{ g \cdot (iy, 0) \mid y \in \omega\}, \quad |g| < \alpha |y|$$

$$\Gamma_{\alpha,\omega}(g_0 \cdot 0) = g_0 \cdot \Gamma_{\alpha,\omega}(0)$$

(in the case of one complex variable these are non tangential angular domains on the upper halfplane). If F, f are functions on D, B

respectively, we say that F converges to f admissibly a.e., if for every fixed α and ω we have

$$\lim_{\substack{z \to u_o \\ z \in \Gamma_{\alpha,\omega}(u_o)}} F(z) = f(u_o)$$

for almost all $u_o \in B$.

Theorem 6. If $f \in L^{\infty}(B)$ and F is the Poisson integral of f, the F converges to f admissibly a.e.

The proof is based on Theorem 4 (e) and on an extension of Lebesgue's differentiation theorem to the group $\mathcal{N}$; it will be contained in a joint article of E. M. Stein and the author in the Annali della Scuola Normale di Pisa.

We should note that for special types of D Theorem 6 can be considerably strengthened. When D is isomorphic with a symmetric domain of tube type, Norman Weiss [21] showed that it is enough to assume $f \in L^p(B)$ $(p > 1)$. This was still strengthened by E. M. Stein (unpublished, as of now) to include the case $p=1$. The case of $f \in L^1(B)$, D isomorphic with a complex ball, was settled by the author. N. Weiss has recently extended these results to include all D isomorphic with classical symmetric domains (for $f \in L^1(B)$).

Theorem 7. If $F \in H^p(D)$ $(1 \leq p \leq \infty)$, then

F is the Poisson integral of a function in $L^p(B)$.

The proof for $p > 1$ is a rather standard argument; it is explicity given in [14] p. 337. The case of $p = 1$ is much harder (is one variable this is called the F. and M. Riesz theorem); it is due to E. M. Stein [20].

Note. From this theorem it follows that, if $F \in H^p(D)$ is the Poisson integral of f, then $\|F_t\|_{L^p(B)} \leq \|f\|_{L^p(B)}$ for all $t \in \Omega$ (immediate from the Remark after Theorem 4). This shows that the map $F \mapsto f$ imbeds the Banach space $H^p(D)$ isometrically into $L^p(B)$ (if $1 \leq p \leq \infty$).

In the case of the one-variable upper halfplane the set of Poisson integrals of all (say, bounded) functions on B can be characterized as the set of (bounded) harmonic functions on D. It would be interesting to have an analogous result in the general case. If D is symmetric, we shall show in sec. 5 that every Poisson integral is harmonic in the sense of symmetric spaces (but not every harmonic function is a Poisson integral). In the case of a general D nothing along these lines is known.

A. Koranyi

5. The Cayley transform of a bounded symmetric domain.

Let D be a bounded symmetric domain imbedded in $\mathcal{C}^+$ in the standard way (sec. 2). We want to describe $c \cdot D$; in the case of the unit disc this amounts to rotating the Riemann sphere by the angle of $\frac{\pi}{2}$ around a horizontal axis, and thereby mapping the unit disc onto the upper halfplane, ($z \mapsto \frac{z + i}{iz + 1}$, which is equivalent to the usual Cayley transform).

A priori it is not clear that $c \cdot D$ makes sense; the action of c on $\mathcal{C}^+$ is defined as $\xi^{-1} c\, \xi$ (and $D = \xi^{-1}(M)$), so we need to know that $\xi(\mathcal{C}^+)$ contains cM . This will turn out to be the case, and we shall see that $c \cdot D$ is always a generalized halfplane.

To compute $c \cdot D$ we use the fact that F_0 is transitive on D (cf. sec. on S as a homogeneous space), so $c \cdot D = (\mathrm{ad}(c)F_0)(c \cdot 0) = (\mathrm{ad}(c)F_0)(iE_\Delta)$. $\mathrm{ad}(c)\, F_0 = K'^* \cdot N^+$ by theorem 3-3 ; we have to study the action of this group on $\mathcal{C}^+$. (By definition of F , ad(c)F is the isotropy group of ad(c)G at $c \cdot iE$; this latter point can be interpreted as a "point at infinity" in $\mathcal{C}^+$.) We know that K'^* , being a subgroup of $K^{\mathbb{C}}$, acts on $\mathcal{C}^+$ by the adjoint representation.

Lemma 1 . $\Omega = K_1^*(E_\Delta)$ is a cone in the real form $\mathcal{U}_1^*$ of $\mathcal{C}_1^+$, self-dual with respect to the restriction of the quadratic form B_τ .

Proof . We show first that $E_\Delta \in \mathcal{U}_1^+$. We know that $E_\Delta \in \mathcal{C}_1^+$; by definition of $\mathcal{U}_1^+$ we have to show only that

$E_\Delta \in ad(c)\mathfrak{g}$. For this we compute $ad(c)^{-1}E_\Delta$; this computation is reduced by strong ortogonality to a computation in the 3-dimensional simple algebra, and the identity

$$\frac{1}{2}\begin{pmatrix}1 & -i\\ -i & 1\end{pmatrix}\begin{pmatrix}0 & 1\\ 0 & 0\end{pmatrix}\begin{pmatrix}1 & i\\ i & 1\end{pmatrix} = \frac{1}{2}\begin{pmatrix}i & 1\\ 1 & -i\end{pmatrix}$$

shows that $ad(c)^{-1}E_\Delta \in \mathfrak{g}$,

Next we note that by theorem 3-2, applied to $\mathfrak{g}_1^{\mathbb{C}}$ instead of $\mathfrak{g}^{\mathbb{C}}$, K_1^* normalizes $\mathfrak{n}_1^+$, i.e. acts on $\mathcal{C}_1^+$ by real linear transformations. It follows $\Omega \subset \mathfrak{n}_1^+$.

The (connected) isotropy subgroup of K_1^* at iE_Δ is $K_1^* \cap F_o = (L_1)_o$; a dimension count now shows that Ω is open in $\mathfrak{n}_1^+$.

We have $ad(c)^2 z_1 = -z_1$ (immediate from Lemma 3-1), hence $z_1 \in \mathfrak{g}_1$. Then $iz_1 \in \mathfrak{k}_1^*$, $\exp ti\, z_1 \in K_1^*$ $(t \in R)$. This is the group of real homotheties; it preserves the set Ω , so Ω is a cone.

$\mathfrak{n}_1^+$ is identified with its dual under B_τ. The group of adjoints of the linear transformations in K_1^* acts transitively on Ω' , by some (relatively easy) results of Vinberg ([20a] , Ch. I, Propositions 9,10). The group of adjoints of K_1^* coincides with K_1^*, since $\mathfrak{k}_1^* = \mathfrak{l}_1 + i\,\mathfrak{g}_1$ and the linear transformations corresponding to $\mathfrak{l}_1$, $i\,\mathfrak{g}_1$ on $\mathfrak{n}_1^+$ are skew-symmetric and symmetric, respectively. Therefore, to show that $\Omega' = \Omega$ it is enough to show that $\Omega' \cap \Omega \neq \emptyset$, by showing e.g. that $E_\Delta \in \Omega'$. To see this we note that $\Omega = ad(K_1^*) \cdot E_\Delta = \exp\,(ad(i\,\mathfrak{g}_1)) \cdot E_\Delta$. For every $V \in i\,\mathfrak{g}_1$, $\exp(ad(V))$ is positive definite on $\mathfrak{n}_1^+$, since ad(V) is symmetric . Therefore $B_\tau(\exp(\mathrm{ad}(V)) \cdot E_\Delta, E_\Delta) > 0$, i. e. $B_\tau(E, E_\Delta) > 0$ for all $E \in \Omega$. This means exactly that $E_\Delta \in \Omega'$, finishing the proof.

Lemma 2 . For all $U \in \mathcal{C}_2^-$, $\mathrm{ad}(c)^2 U = -[U, iE_\Delta]$.

Proof. First we show that, restricted to $\mathcal{C}_2^{\mathbb{C}} + \mathcal{V}_2^{\mathbb{C}}$, $\mathrm{ad}(c)^2 = i\,\mathrm{ad}(X)$. For this we have to note that $\mathcal{C}_2^{\mathbb{C}} + \mathcal{V}_2^{\mathbb{C}}$ is the union of irreducible representation spaces on which ad(c) has order 8 . Hence $\mathrm{ad}(c)^2 = \exp(\mathrm{ad}(\frac{\pi}{2} iX))$ must have eigenvalues $\pm i$ on it which in turn implies that $\mathrm{ad}(\frac{\pi}{2} iX)$ must have eigenvalues $\pm i$ (since $e^{\pm \frac{\pi}{2} i} = \pm i$), and the assertion follows .

To prove the Lemma, let $U \subset \mathcal{C}_2^-$. We have

$$\mathrm{ad}(c)^2 U = -[U, X] = -[U, iE_\Delta] - [U, iE_{-\Delta}] = -[U, iE_\Delta]$$

(using at the last step that $\mathcal{C}^-$ is abelian).

For a linear transformation T on $\mathcal{G}^{\mathbb{C}}$ we denote by T^* its adjoint with respect to B_τ.

Lemma 3 . For $U, V \in \mathcal{C}_2^+$ we have

$$\Phi(U, V) = \frac{1}{2}\mathrm{ad}(U)\,\mathrm{ad}(V)^* \; E_\Delta = -\frac{i}{2}\left[U, \mathrm{ad}(c)^2\, \tau(V)\right].$$

Φ is a function $\mathcal{C}_2^+ \times \mathcal{C}_2^+ \longrightarrow \mathcal{C}_1^+$ with the properties :

(i) Φ is complex-linear in the first argument,

(ii) Φ is Hermitian with respect to $\mathcal{U}_1^+$,

(iii) for all $k \in K'^*$,

$\Phi(\mathrm{ad}(k)U, \mathrm{ad}(k)V) = \mathrm{ad}(k)\Phi(U,V)$,

(iv) $\Phi(U,U) \in \overline{\Omega}$ for all $U \in \mathcal{C}_2^+$, and $\Phi(U,U) = 0$ implies $U = 0$.

Proof. Using $\tau(V) \in \mathcal{C}_2^-$, Lemma 2, and the easily checked fact that $\mathrm{ad}(V)^* = -\mathrm{ad}(\tau V)$, we obtain $(\mathrm{ad}(c)^2\tau V = -[\tau V, iE_\Delta] = -\mathrm{ad}(\tau V)iE_\Delta = \mathrm{ad}(V)^* iE_\Delta$, proving the first equality. It is trivial to check that $\Phi(U,V)$ is in the (+i)-eigenspace of $\mathrm{ad}(z_1)$, so in $\mathcal{C}_1^+$. Property (i) of Φ is also trivial.

To show (ii), let μ denote complex conjugation in $\mathcal{C}_1^+$ with respect to $\mathfrak{n}_1^+$. Since $\mathfrak{n}_1^+ = \mathcal{C}_1^+ \cap \mathrm{ad}(c)\mathfrak{g}$, μ is the restriction of the complex conjugation of $\mathfrak{g}^{\mathbb{C}}$ with respect to $\mathrm{ad}(c)\mathfrak{g}$, i.e. of $\mathrm{ad}(c)\sigma\tau\,\mathrm{ad}(c)^{-1} = \sigma\,\mathrm{ad}(c)^{-1}\,\tau\,\mathrm{ad}(c)^{-2}$.

Now let $U, V \in \mathcal{C}_2^+$. We have

$$\sigma\tau\,\mathrm{ad}(c)^{-2}U = -\sigma\tau\,\mathrm{ad}(c)^2U = -\sigma\,\mathrm{ad}(c)^2\tau U = -\mathrm{ad}(c)^2\tau U$$

$$\sigma\tau\,\mathrm{ad}(c)^{-2}(\mathrm{ad}(c)^2\tau V) = \sigma V = -V$$

and hence

$$\mu\Phi(U,V) = \sigma\tau\,\mathrm{ad}(c)^{-2}\Phi(U,V) = \frac{1}{2}\left[-\mathrm{ad}(c)^2\tau U, -V\right] = \Phi(V,U).$$

To prove (iii) we note that $\mathrm{ad}(c)^2$ and τ are both trivial on $\mathfrak{l}$ and equal to $-I$ on $i\mathfrak{g}_1$. Hence their product is trivial on $\mathfrak{K}'^*$, which means that it commutes with every element of $\mathrm{ad}(K'^*)$. From the second form of the definition of Φ the statement is now obvious.

To show $\Phi(U,U) \in \overline{\Omega}$ it suffices to show $B_\tau(\Phi(U,U),V) \geqq 0$ for all $V \in \Omega$ (since Ω is self-dual).

Given $V \in \Omega$, we choose $k \in K'^*$ such that $ad(k)V = E_\Delta$. Since $ad(k)$ is unitary, by (iii) we have $B_\tau(\Phi(U,U),V) = B_\tau(\Phi(U',U'),E_\Delta)$ where $U' = ad(k)U$. This is further equal to

$\frac{1}{2}B_\tau(ad(U')ad(U')^* E_\Delta, E_\Delta)$, which is $\geqq 0$ by the positive definiteness of B_τ.

If $\Phi(U, U) = 0$, then $ad(U)^* E_\Delta = 0$ by the last formula. This means that $-ad(\tau U)E_\Delta = 0$. Since $\tau U \in \mathcal{G}_2^-$, it follows that $[\tau U, Y] = 0$. By Lemma 3-14 then $\tau U = 0$, whence $U = 0$, finishing the proof.

Lemma 4. $I - ad(c)^2\tau : \mathcal{G}_2^+ \rightarrow \mathcal{W}_2^+$ is a linear isomorphism.

Proof. Let $V \in \mathcal{G}_2^+$; then $\tau V \in \mathcal{G}_2^-$ and $ad(c)^2 \tau V \in \mathcal{Y}_2^+$. Considering the definition of $\mathcal{W}_2^+$, we have to show only that $V - ad(c)^2 \tau V \in ad(c)\mathcal{Y}$ (then the map will also be surjective, by counting dimensions). To show this we apply the conjugation with respect to $ad(c)\mathcal{Y}$, which was computed above to be $\sigma\tau\, ad(c)^{-2}$. The same computation as in Lemma 3 (ii) shows that it leaves $V - ad(c)^2\tau V$ fixed.

From now on, given $E \in \mathcal{G}^+$, we shall denote its projection onto $\mathcal{G}_1^+$ and $\mathcal{G}_2^+$ by E_1, E_2. So $E = E_1 + E_2$.

Theorem 1. Every $g \in N^+$ is of the form $g = \exp(U + (I - ad(c)^2\tau)V)$ for some $U \in \mathcal{W}_1^+$, $V \in \mathcal{G}_2^+$. It acts on $\mathcal{G}^+$ by

$$g \cdot E = E + U + V + 2i\Phi(E_2, V) + i\Phi(V, V).$$

Proof. The first statement follows from Lemma 4. Applying the Campbell-Hausdorff formula, we have

$$g = \exp(U + (I - \mathrm{ad}(c)^2 \tau) V) =$$
$$= \exp(U) \cdot \exp\left(\tfrac{1}{2}\left[V,\ \mathrm{ad}(c)^2 \tau V\right]\right) \cdot \exp(V) \cdot \exp(-\mathrm{ad}(c)^2 \tau V),$$

since all other brackets vanish by Lemma 3-14. The action of each factor on $\mathcal{C}^*$ is easy to compute. We have $-\mathrm{ad}(c)^2 \tau V \in \mathfrak{g}_2^+ \subset \mathfrak{K}^{\mathbb{C}}$ so the last factor acts by the adjoint representation; exponentiating we get

$$\exp(-\mathrm{ad}(c)^2 \tau V) \cdot E = E - \left[\mathrm{ad}(c)^2 \tau V,\ E\right] =$$
$$= E + \left[E_2,\ \mathrm{ad}(c)^2 \tau V\right]$$

since all other brackets vanish, again by Lemma 3-14. The other three factors in the factoring of g belong to P^+; therefore they act on $\mathcal{C}^+$ by translations. Using now the definition of Φ, we obtain the statement.

Theorem 2. If D is the bounded symmetric domain constructed in sec. 2, then its Cayley transform, $c \cdot D$ is the generalized halfplane $\{E \mid \mathrm{Im}\, E_1 - \Phi(E_2, E_2) \in \Omega\}$.

Proof. For this generalized halfplane the role of $\mathcal{N}$ (in the sense of sec. 4) is played by N^+, and $G_0(\Omega, \Phi) = K'^*$ ($K'^* \subset G_0(\Omega, \Phi)$ follows from Lemma 3 (iii); equality can be shown by using that Ω is a symmetric space, but is unimportant). Our generalized halfplane is affine homogeneous, since K'^* acts transitively on Ω. It contains iE_Δ, and therefore equals the orbit $K'^* \cdot N^+(iE_\Delta) = (\mathrm{ad}(c)F_0)(iE_\Delta)$. By the remarks at the beginning of this section, this orbit is $c \cdot D$.

Theorem 3. Let $B = \{ E \mid \operatorname{Im} E_1 - \Phi(E_2, E_2) = 0 \}$ be the distinguished boundary of $c \cdot D$. Then $c^{-1} \cdot B$ is a dense open subset of S, the Bergman-Shilov boundary of D.

Proof. We have $B = (N^+ \cdot K'^*) \cdot 0 = (\operatorname{ad}(c) F_0) \cdot 0$. Then $c^{-1} \cdot B = F_0 \cdot (-iE_\Delta)$, since $c^{-1} \cdot 0 = -iE_\Delta$ as it is easy to check. It follows by a dimension count that $c^{-1} \cdot B$ is open in S. Assume it is not dense. Then there exists a point u_0 in the inferior of $S - c^{-1} \cdot B$, and a function f holomorphic on D, continuous on $\overline{D}$, such that $|f|$ assumes its maximum at u_0 and nowhere else. Then $g = f \circ c^{-1}$ is a bounded holomorphic function on $c \cdot D$, continuous on $c \cdot \overline{D}$, and such that $\sup_{c \cdot D} |g| > \sup_{B} |g|$.

This is impossible by the results of sec. 4.

Explicit formulas and harmonocity of $P(u, z)$.

We want to find more explicit formulas than those given by Theorem 4-3 for the Szegö and Poisson kernels of $c \cdot D$. We will also determine the Bergman kernel (up to constant multiple), which is necessary for the proof that $P(u, z)$ (and therefore every Poisson integral) is harmonic.

We use the norm function $N : \Omega \longrightarrow R$ and the Γ- function of Koecher, which are defined by

$$\Gamma^*(s) = N(y)^s \int_\Omega e^{-B_\tau (y, t)} N(t)^{s-1} dt$$

$$N(E_\Delta) = 1 .$$

By a simple change of variable one sees that, for $g \in K'^{*}$,

$$N(gy) = (\det g)\, N(y)$$

where $\det g$ is the determinant of $ad(g)|_{\mathfrak{n}_1^+}$. The same change of variable shows that everything is well defined. We note that, as Gindikin showed, cf. [6], [15],

$$\Gamma^{*}(s) = (2\pi)^{\frac{n_1 - \ell}{2}} \prod_{j=0}^{\ell - 1} \Gamma\left(\frac{n_1}{\ell} s - j \frac{n_1 - \ell}{(\ell - 1)}\right)$$

where $n_1 = \dim \mathfrak{n}_1^+$ and ℓ is the rank.

<u>Theorem</u> 4. The Szegö kernel of $c \cdot D$ is

$$S(z, w) = \frac{1}{2^{n_1} \pi^{n}} \Gamma^{*}\left(1 + \frac{n_2}{n_1}\right) N(\rho(z, w))^{-1 - \frac{n_2}{n_1}}.$$

<u>Proof</u>. We look at $\det B_\lambda$ in the formula of Theorem 4-3. Let $g \in K_1^{*}$, then a trivial computation shows that $\det B_{g\lambda} = (\det B_\lambda) \cdot (\det ad(g)|_{\mathscr{C}_2^+})^2$. The latter factor must be equal to $(\det g)^k$ for some k (where $\det g$ is, as above, relative to $\mathfrak{n}_1^+$), since all 1-dimensional representations of K_1^{*} are of this form (recall that K_1, hence also K_1^{*} is reductive with 1-dimensional center). $\det B_\lambda$ is homogeneous of degree n_2, hence $k = {}^{n_2}/n_1$, and $\det B_\lambda = cN(\lambda)^{n_2/n_1}$. The constant c

can be computed to be 2^{n_2} by verifying that $B_{E_\Delta} = 2B_\tau$ (this depends on some identities for roots; it is Lemma 5.2 in [14]). The statement follows now immediately from the definition of the norm function.

Note . A more explicit way of computing the norm will be indicated in sec. 6 .

Lemma 5 . The Bergman kernel function of $c \cdot D$ is

$$K(z, w) = cN(\rho(z, w))^{-2-n_2/n_1} .$$

Proof . From the general theory of the Bergman kernel one knows (e.g. [8] p. 293) that (i) $K(z,w)$ is holomorphic in the first, antiholomorphic in the second variable, (ii) $K(gz, gz) \left| \frac{\partial(gz)}{\partial(z)} \right|^2 = K(z,z)$ for all automorphisms g of the domain. For a homogeneous domain these properties clearly determine $K(z,w)$ up to a constant factor ; it is even enough to consider g in some fixed transitive subgroup of the group of all automorphisms.

It is obvious that our formula fulfills (i). We check (ii) for the transitive group $ad(c)F_o = K'^* \cdot N^+ \cdot N^+$ leaves $\rho(z,w)$ invariant, and also has Jacobian 1 everywhere on D . So (ii) holds for $g \in N^+$. If $g \in K'^*$, $\frac{\partial(gz)}{\partial(z)}$ will be equal to the determinant of $ad(g)|_{\mathcal{G}^+}$ (cf. Lemma 3 (iii)). By the computation in the proof of theorem 4 , this is equal to $(\det g)^{-2-n_2/n_1}$ with $\det g$ referring to the action of g on $\mathcal{U}_1^+$

Now $N(\rho(gz, gw))^{-2-n_2/n_1} = N(g\rho(z,w))^{-2-n_2/n_1} =$

$= (\det g)^{-2-n_2/n_1} N(\rho(z,w))^{-2-n_2/n_1}$ finishing the proof.

Note . With a considerable amount of extra trouble one finds that

$$c = \frac{\Gamma^*(2 + {}^{n_2}/n_1)}{\pi^n \, \Gamma^*(1)}$$

(cf. [14] p. 346 and references given there); this is also the reciprocal of the Euclidean volume of the domain D .

Theorem 5 . For every fixed $u \in B$, $P(u,z)$ is a harmonic function on $c \cdot D$. (Harmonic means that it is annihilated by all G-invariant differential operators , or what is the same, it has the mean value property with respect to the orbits of the isotropy group at every point.)

Proof . The following proof, which we shall only sketch, is rather roundabout, but seems to be the only one known (see [14] § 3 for more details).

The crucial point is to prove that

$$(*) \qquad P(gu, gz) \frac{d\beta(gu)}{d\beta(u)} = P(u,z)$$

for all $g \in ad(c)G$, $z \in c \cdot D$, and $u \in B$ such that gu is still in B (these u form a dense open set in B for every fixed g). Once this is known, one can deduce the mean value relation

$$\int_{K_w} P(u, kz)\, dk = P(u, w) = P_w(u)$$

(K_w being the isotropy group at $w \in c \cdot D$), by noticing that P_w is characterized by the properties

$$\text{(i)} \quad \int_B P_w \, d\beta = 1 , \qquad \text{(ii)} \quad P_w(hu) = P_w(u)\left(\frac{d\beta(hu)}{d\beta(u)}\right)^{-1}$$

for $h \in K_w$ (in virtue of the transitivity of K_w on B). Using the invariance of the Haar measure, it is trivial to check that the left hand side also satisfies (i) and (ii).

To prove (✱) we proceed as follows. By theorem 4 and Lemma 5, $S(z, w) = cK(z, w)^k$ (the value of k does not matter). One knows how K transforms under automorphisms; one deduces that

$$S(gz, gw)A_g(z)\overline{A_g(w)} = S(z,w) \quad \text{for all} \quad z, w \in c \cdot D, \quad g \in ad(c)G,$$

with $A_g(z) = \left(\frac{\partial(gz)}{\partial(z)}\right)^k$ (one chooses a holomorphic branch). By definition of $P(u, z)$ it follows that

$$P(gu, gz) \left| A_g(u) \right|^2 = P(u, z)$$

for $u \in B$ such that $gu \in B$.

One shows next that, for $g \in ad(c)G$,

$U_g : H^2(c \cdot D) \longrightarrow H^2(c \cdot D)$ defined by $(U_g f)(z) = f(g^{-1}z)A_{g^{-1}}(z)$

is a unitary transformation. In fact, it is trivial to check $(U_g S_z, U_g S_w) = (S_z, S_w)$, and the statement follows from the fact that the S_z $(z \in c \cdot D)$ are dense in $H^2(c \cdot D)$. Using this, we have for all $f \in H^2(c \cdot D)$,

$$\int_B |f(gu)|^2 \frac{d\beta(gu)}{d\beta(u)} \, d\beta(u) = \|f\|^2 =$$

$$= \|U_{g^{-1}} f\|^2 = \int_B |f(gu)|^2 \, |A_g(u)|^2 \, d\beta(u).$$

One sees easily that there are enough functions f in H^2 so that this can be true for all f only if

$$\frac{d\beta(gu)}{d\beta(u)} = |A_g(u)|^2 , \quad \text{which}$$

finishes the proof of $(*)$.

<u>Note</u> . It is now not difficult to see that, given a bounded continuous function f on B , there exists a unique function F harmonic on $c \cdot D$ and continuous on $c \cdot \overline{D}$, such that $F|_B = f$.

6. The Poisson integral on the bounded domain D .

In this section we will translate the results of sec. 5 from $c \cdot D$ to D; we will also find rather explicit formulas for the Szego and Poisson kernels of D . (It is interesting to note that the only known way of getting to these results [14] is by passing through $c \cdot D$. This seems to indicate that in some sense a generalized halfplane is a more elementary object than a bounded symmetric domain in the standard (bounded) realization . It is also true that all bounded homogeneous domains can be realized as (affine-homogeneous) generalized halfplanes (result of Vinberg, Gindikin and Pjateckij-Shapiro), but it seems that they do not in general have a canonical bounded realization.)

We denote by μ the normalized K-invariant measure on S, the Bergman-Shilov boundary of D (in fact we may just think of S as $S = K(iE_\Delta) = G(iE_\Delta)$; the results of this section will show that it is the Bergman-Shilov boundary). $L^p(S)$ will be the L^p-space with respect to μ $(0 < p \leqq \infty)$. Given $f : D \to \mathbb{C}$ and $0 < r < 1$, we define $f_r : D \to \mathbb{C}$ by $f_r(z) = f(rz)$. Generalizing a classical definition in the unit disc, we define $H^p(D)$ $(0 < p < \infty)$ as the space of holomorphic functions f on D such that

$$\|f\|_p = \sup_{0<r<1} \|f_r/_S\|_{L^p(S)} < \infty$$

and $H^\infty(D)$ as the space of bounded holomorphic functions .

For the sake of brevity in this section we denote E_Δ by e . We define the Poisson kernel of D by

$$\mathcal{P}_z(u) = \mathcal{P}(u,z) = \frac{P(cu, cz)}{P(cu, ie)}$$

for $z \in D$, $u \in c^{-1} \cdot B$. We extend its definition to $D \times S$ by the following lemma.

Lemma 1. For every fixed $z \in D$, $\mathcal{P}_z$ has a unique continuous extension to S. For all $k \in K$, $\mathcal{P}(ku, kz) = \mathcal{P}(u, z)$.

Proof. $c^{-1} \cdot B$ is dense open in S and K is transitive on S; so everything is proved if we prove the second statement for $u \in c^{-1} \cdot B$. For this let us use the notation k^c for ckc^{-1}; we know that $k^c \cdot ie = ckc^{-1} \cdot (c \cdot 0) = c \cdot 0 = ie$ for $k \in K$. Using (*) from the proof of theorem 5-5, we find

$$\mathcal{P}(ku, kz) = \frac{P(k^c \cdot cu,\ k^c \cdot cz)}{P(k^c \cdot cu,\ k^c \cdot ie)} = \frac{P(cu,\ cz)}{P(cu,\ ie)} = \mathcal{P}(u, z)$$

finishing the proof.

Lemma 2. For all $u \in c^{-1} \cdot B$,

$$\frac{d\beta(cu)}{d\mu(u)} = P_{ie}(cu)^{-1} .$$

Proof. By (*) in Theorem 5-5, $P_{ie}(cu)\, d\beta(cu)$ is a normalized K-invariant measure on $c^{-1} \cdot B$. Hence it must be equal to $d\mu(u)$.

Remark . Using this, one computes easily that $\mathcal{P}(gu, gz)\frac{d\mu(gu)}{d\mu(u)} = \mathcal{P}(u,z)$, which is equivalent with the fact that $\mathcal{P}(u,z)$ is harmonic (cf. Theorem 5-5). But, of course, one knows even without this argument that $\mathcal{P}(u,z)$ is harmonic: For $u \in c^{-1} \cdot B$ this is immediate from the definition of $\mathcal{P}(u,z)$, and for any $u \in S$ it follows using Lemma 1.

Theorem 1.

(i) $\mathcal{P}(u,z) > 0$ for all $u \in S$, $z \in D$.

(ii) For all fixed $z \in D$, $\mathcal{P}_z$ is continuous on S (even ailytic).

(iii) $\int_S \mathcal{P}_z \, d\mu = 1$, for all $z \in D$.

(iv) For every $u_0 \in S$ and every neighborhood N of u_0 in S, $\lim_{z \to u_0} \int_{u \notin N} \mathcal{P}(u,z)\, d\mu(u) = 0$.

(v) Defining, for functions $f : S \longrightarrow \mathbb{C}$, $\tilde{f}(k) = f(k \cdot c)$ and defining, for $0 < r < 1$, $\tilde{\mathcal{P}}_r(k) = \mathcal{P}(e, k \cdot re)$ $(k \in K)$, we have for the Poisson integral $F(z) = \int_S f \, \mathcal{P}_z \, d\mu$,

$$\tilde{F}_r = \tilde{f} * \tilde{\mathcal{P}}_r$$

(convolution on K).

(vi) $P(u,z)$ for every fixed $u \in S$ is a harmonic function on D.

Proof . (i) We show $\mathcal{P}_z > 0$ everywhere on S by showing $P_z > 0$ everywhere on B. Using $(*)$

(Theorem 5-5) if is enough to show that P_{ie} 0 everywhere on B (in fact we should have shown this when we defined $\mathcal{P}(u, z)$...) Assume $P_{ie}(u_o) = 0$ for some $u_o \in B$. Then, by (✱), $P_{ie}(k^c \cdot u_o) = 0$ for all $k \in K$, i.e $P_{ie} \equiv 0$. Then $S_{ie} \equiv 0$ which by Theorem 4-3 is absurd.

(ii) is immediate from Theorem 4-4 (b) and Lemma 1. To prove (iii) we use Lemma 2 and Theorem 4-4 (c) to get

$$\int_S \mathcal{P}_z \, d\mu = \int_S \frac{P(cu, cz)}{P(cu, ie)} \, d\mu(u) =$$

$$= \int_B P(cu, cz) \, d\beta(cu) = 1 .$$

The proof of (iv) is analogous.

(v) Is immediate by the equalities

$$\widetilde{F}_r(h) = F_r(he) = F(h \cdot re) =$$

$$= \int_S f(u) \, \mathcal{P}(u, h \cdot re) d\mu(u) =$$

$$= \int_K \widetilde{f}(k) \, \mathcal{P}(-e, k^{-1} h \cdot re) \, dk =$$

$$= (\widetilde{f} * \mathcal{P}_r)(h) .$$

(vi) Was already proved in the remarks preceding the statement of the theorem.

Corollaries. The exact analogues (with S written in place of B) of Theorems 4-5 and 4-7 follow immediately from Theorem 1. (To get the analogue of 4-7 in the case $p = 1$ one has to argue a little more: In order to make sure that the boundary function is a function in $L^1(B)$, and not only a measure, one can use the analogous statement in Theorem 4-7 and the Cayley transform c.) It is also clear that, for $1 \leq p \leq \infty$, $H^p(D)$ can be regarded as a subspace of $L^p(S)$.

Theorem 4-6 could also be translated to our D, but this is not too interesting, since the results quoted after Theorem 4-6 seem to indicate that for the symmetric domain D considerably stronger theorems hold.

Now we take a closer look at $H^2(D)$ making use of a simplification of the method of [14] suggested by N. Weiss. For every $f \in H^2(c \cdot D)$ we define

$$Tf : D \cup S \longrightarrow C \qquad \text{by}$$

$$(Tf)(z) = S(ie, ie)^{1/2} S_{ie}(cz)^{-1} f(cz).$$

Tf is well defined on D, and is defined a.e. on S, since by Theorem 4-7 f is defined a.e. on B.
(Tf is holomorphic on D since we know that $S_{ie}(cz) \neq 0$; in fact, in the contrary case we would have $S_{ie}(k^c cz) = 0$ for all $k \in K$ by the transformation rule of S under automorphisms (cf. proof of Theorem 5-5), and the mean value property would imply $S_{ie}(ie) = 0$, which is absurd.)

We also define $\mathcal{S} : D \times D \longrightarrow C$ by

$$\mathcal{S}(z, w) = \frac{S(ie, ie) \; S(cz, cw)}{S(cz, ie) \; S(ie, cw)}$$

and $\mathcal{S}_w$ by $\mathcal{S}_w(z) = \mathcal{S}(z, w)$. It is immediate to check

$$\mathcal{S}_w = S(ie, ie)^{1/2} S(ie, cw)^{-1} T S_{cw}$$

and (noting that $\mathcal{S}$ extends to $\bar{D} \times D$),

$$\mathcal{P}(u, z) = \frac{|\mathcal{S}(u, z)|^2}{\mathcal{S}(z, z)} .$$

Theorem 2. T is a Hilbert space isomorphism of $H^2(c \cdot D)$ onto $H^2(D)$. $\mathcal{S}$ is the Szegö kernel of $H^2(D)$.

Proof. (a) For every $f, g \in H^2(c\,D)$ we have $(Tf, Tg)_{L^2(S)} = (f, g)$. In fact, by definition of T and by Lemma 2,

$$(Tf, Tg)_{L^2(S)} = S(ie, ie) \int_S | S_{ie}(cu) |^{-2} f(cu) \overline{g(cu)} \, d\mu(u) =$$

$$= \int_S f(cu) \overline{g(cu)} P_{ie}(cu)^{-1} \, d\mu(u) =$$

$$= \int_B f(v) \overline{g(v)} \, d\beta(v) = (f, g) .$$

(b) For every $f \in H^2(c \cdot D)$ we have $(Tf, \mathcal{S}_z)_{L^2(S)} = (Tf)(z)$. This is clear by using our previous formula for $\mathcal{S}_z$ and applying (a).

(c) For every $f \in H^2(c \cdot D)$, $(Tf, \mathcal{P}_z)_{L^2(S)} = (Tf)(z)$. To see this, we write $(Tf, \mathcal{P}_z) = \mathcal{S}(z, z)^{-1}((Tf)\mathcal{S}_z, \mathcal{S}_z)$, then note that $(Tf)\mathcal{S}_z$ is the image under T of an element of $H^2(c \cdot D)$, so (b) can be applied.
Continueing the equality we get $= \mathcal{S}(z, z)^{-1} \cdot (Tf)(z) \cdot \mathcal{S}_z(z) = (Tf)(z)$.

(d) $Tf \in H^2(D)$ for all $f \in H^2(c \cdot D)$. This is immediate from (c) and Theorem 1 (v).

(e) T is an isometric map by (a), and is surjective since the image under T clearly contains all functions holomorphic on $\overline{D}$, and every $f \in H^2(D)$ can be approximate (in $H^2(D)$) by such functions (e.g. since $f = \lim_{r \to 1} f_r$).

(f) $\mathcal{S}$ is the reproducing kernel of $H^2(D)$: Since we know that every element of $H^2(D)$ is of the form Tf, this is now obvious from (b).

Remarks . One can easily show ([14], p. 345) that the Bergman kernel of D is again constant times a power of the Szego kernel. It is also easy to check that, for all $g \in G$,

$$\mathcal{S}(gz, gw) \mathcal{A}_g(z) \overline{\mathcal{A}_g(w)} = \mathcal{S}(z, w)$$

where $\mathcal{A}_g(z)$ is a power of the Jacobian determinant of g acting on $\mathcal{C}^+$. In particular, since $k \in K$ acts by unitary linear transformations,

$$\mathcal{S}(kz, kw) = \mathcal{S}(z, w) .$$

(The last equality can also be proved directly in an elementary way, by noticing that $f \mapsto f \circ k$ is an automorphism of $H^2(D)$.)

A. Koranyi

Explicit formulas. We shall get a formula for the Szego kernel (and therefore also for the Poisson kernel) of D with a high degree of explicitness, which makes the precise computation of this kernel in every concrete case very easy. We shall also indicate how one can compute such constants as the volume of D or B. (It makes good sense to talk about these volumes, since according to the conventions and results of sec. 2, the size and shape of D are uniquely determined by the conditions that the isotropy group at 0 acts by unitary linear transformations and that the largest inscribed sphere in D has radius 1.) For all this we have to make some further computations in the spirit of sec. 5.

By Lemma 3-1, $\mathrm{ad}(c)^2 H_\alpha = -H_\alpha$ for $\alpha \in \Delta$. Hence $\mathfrak{f}^- \subset \mathfrak{g}$. Defining $\mathfrak{f}^* = i\mathfrak{f}^-$, we have $\mathfrak{f}^* \subset i\mathfrak{g} \subset \mathfrak{K}_1^*$. We denote the corresponding analytic subgroup of G^C by H^*.

(It is easy to see that $\mathfrak{f}^*$ is a Cartan subalgebra of the pair $(\mathfrak{K}_1^*, \mathfrak{l}_1)$. Therefore $K_1^* = (L_1)_0 H^* (L_1)_0$ and $\Omega = (L_1)_0 H^*(e)$, a useful fact for making computations on Ω).

H^*, being a subgroup of K^c, acts on $\mathfrak{p}^+$ by the adjoint representation. A typical element of H^* is of the form $g = \exp \sum_{\alpha\in\Delta} t_\alpha H_\alpha$, we have $g \cdot e = \sum_{\alpha\in\Delta} e^{t_\alpha(H_\alpha)} E_\alpha = \sum_{\alpha\in\Delta} e^{2t_\alpha} E_\alpha$ by strong orthogonality of Δ. It follows that

$$H^*(e) = \left\{ \sum_{\alpha\in\Delta} y_\alpha E_\alpha \,\middle|\, 0 < y_\alpha < \infty \right\}.$$

Lemma 3. For $y = \sum_{\alpha\in\Delta} y_\alpha E_\alpha$ $(0 < y_\alpha < \infty)$ we have

$$N(y) = \prod_{\alpha\in\Delta} y_\alpha^{-n_1/\ell}$$

(where N is the norm function of Ω, ℓ = rank D).

Proof . By the remark just made, $y = g\,e$ with $g = \exp \sum_{\Delta} (\log \frac{y_\alpha}{2})\, H_\alpha$. We know that $N(y) = \det(\mathrm{ad}(g)\,|\,\mathfrak{n}_1^+\,|)$; this is now trivial to compute since g acts diagonally with respect to the basis $\{E_\beta\}$ $(\beta \in \Phi)$:

$$g \cdot E_\beta = \prod_{\alpha \in \Delta} y_\alpha^{\frac{\langle\alpha,\beta\rangle}{2}} E_\beta .$$

Hence $N(y) = \det g = \prod_{\Delta} y_\alpha^{\langle\alpha,\rho\rangle}$, where $\rho = \frac{1}{2}\sum_{\beta\in\Phi} \beta$. By Lemma 3-5, $\langle\alpha,\rho\rangle$ is the same for each $\alpha \in \Delta$. Since N must be a homogeneous function of degree n_1, it follows that $\langle\alpha,\rho\rangle = n_1/\ell$.

Lemma 4 . For $z_o = i \sum_{\Delta} r_\alpha E_\alpha$ $(-1 < r_\alpha < 1)$ we have $c \cdot z_o = i \sum_{\Delta} y_\alpha E_\alpha$ with $y_\alpha = \frac{1 + r_\alpha}{1 - r_\alpha}$ $(\alpha \in \Delta)$.

Proof . A direct computation gives

$$\xi(cz_o) = (\exp \sum_{\Delta} i\, \frac{\pi}{4} X_\alpha)(\exp i \sum_{\Delta} r_\alpha E_\alpha) \cdot X =$$

$$= \exp (i \sum_{\Delta} \frac{1 + r_\alpha}{1 - r_\alpha} E_\alpha) \cdot X ,$$

where we had to use strong orthogonality and the identity

$$\frac{1}{\sqrt{2}} \begin{pmatrix} 1 & i \\ i & 1 \end{pmatrix} \begin{pmatrix} 1 & ir \\ 0 & 1 \end{pmatrix} =$$

$$= \begin{pmatrix} 1 & i\dfrac{1+r}{1-r} \\ 0 & 1 \end{pmatrix} \begin{pmatrix} \dfrac{\sqrt{2}}{1-r} & 0 \\ 0 & \dfrac{1-r}{\sqrt{2}} \end{pmatrix} \begin{pmatrix} 1 & 0 \\ \dfrac{i}{1-r} & 1 \end{pmatrix}$$

Theorem 3 . For $z_o = i \sum_{\Delta} r_\alpha E_\alpha$ $(-1 < r_\alpha < 1)$ we have $\mathcal{S}(z_o, z_o) = \prod_{\Delta} (1 - r_\alpha^2)^{n/\ell}$ (where $n = n_1 + n_2 = \dim_{\mathbb{C}} D$).

Proof . Immediate from the definition of $\mathcal{S}$, Theorem 5 - 4 and Lemma 4 .

Example . To illustrate how one gets a precise formula for $\mathcal{S}$ in concrete cases, let us consider the domain D formed by complex $p \times q$ matrices $(p \geq q)$ z such that $I - z^* z$ is positive definite . This is a symmetric domain in standard realization ; the connected isotropy group K at 0 consists of the maps $z \mapsto uzv$ where u, v are ($p \times p$ resp. $q \times q$) unitary matrices . (This example is studied in detail in [18] ; it serves very well to illustrate all that has been done in these lectures .) We have $n = pq$, $\ell = q$, and it is easy to check that by an element of K every z can be transformed into

$$z_o = i \begin{pmatrix} r_1 & & \\ & \ddots & \\ & & r_q \\ & 0 & \end{pmatrix}$$

wich corresponds to the situation in Theorem 3 . The question is how to rewrite the expression in such a form that it be invariant under K, and it is not difficult to notice that this can be done in a unique way by setting $\mathcal{S}(z, z) = \det(I - z^* z)^{-p}$. Since $\mathcal{S}$ is holomorphic in the first, antiholomorphic in the second variable, we must have $\mathcal{S}(z, w) = \det(I - w^* z)^{-p}$.

Remarks. We have been using the normalized K-invariant measure μ on S; one might want to use the volume induced by the Euclidean structure of $\mathcal{C}^+$ instead. Since K acts by unitary transformations, these measures are proportional, the proportionality constant being the Euclidean volume of S. It is easy to check that the Jacobian of c at, say, $-ie$ is 2^{-n} ([14] Lemma 5-6), and then making use of the precise value of the constant in Theorem 5-4 one finds that

$$\operatorname{vol} S = \frac{2^{n_1} \pi^n}{\Gamma^*(1 + n_2/n_1)} .$$

The same kind of computation can be performed with the Bergman kernel of D ([14], sec. 5).

Using Lemma 5-5 one finds an expression analogous to Theorem 3, and associated with it the exact value of the volume of D (cf. Note after Lemma 5-5).

A. Koranyi

References

[1] W. L. Baily and A. Borel, Compactification of arithmetic quotients of bounded symmetric domains, Ann. of Math. 84 (1966), 442-528.

[2] S. Bochner, Group invariance of Cauchy's formula in several variables, Ann. of Math. 45 (1944), 686-707 .

[3] H. Braun and M. Koecher, Jordan-Algebren, Springer 1966 .

[4] E. Cartan, Sur les domaines bornés homogènes de l'espace de n variables, Abh. Math. Sem. Hamburg, 11 (1935), 116-162 .

[5] C. Chevalley, Lie groups, Vol. I, Princeton University Press, 1946.

[6] S. G. Gindikin, Analysis in homogeneous domains , Uspekhi Mat. Nauk 19 (1964), 3-92 (in Russian).

[7] Harish-Chandra, Representations of semi-simple Lie groups VI, Amer. J. Math. 78(1956), 564-628 .

[8] S. Helgason, Differential geometry and symmetric spaces, Academic Press, 1962.

[9] Ch. Hertneck, Positivitätsbereiche und Jordan-Strukturen, Math. Ann. 146 (1962), 433-455.

[10] U. Hirzebruch, Über Jordan-Algebren und beschränkte symmetrische Gebiete, Math. Z. 94 (1966), 387-390 .

[11] G. Hochschild, The structure of Lie groups, Holden-Day , 1965.

[12] L. K. Hua, Harmonic analysis of functions of several complex variables in the classical domains, Amer. Math. Soc. 1963.

[13] N. Jacobson, Lie algebras, Interscience, 1962 .

[14] A. Koranyi, The Poisson integral for generalized half-planes and bounded symmetric domains, Ann. of Math. 82 (1965), 332-350.

[15] A. Koranyi, Analytic invariants of bounded symmetric domains, to appear in Proc. Amer. Math. Soc.

[16] A. Koranyi and J. A. Wolf, Realization of Hermitian symmetric domains as generalized half-planes, Ann. of Math. 81 (1965), 265-288.

[17] C. C. Moore, Compactifications of symmetric spaces II. The Cartan domains, Amer. J. Math. 86 (1964), 358-378 .

[18] I. I. Pyateckii-Shapiro, Geometry of classical domains and automorphic functions, Fizmatgiz 1961 (in Russian).

[19] I. Satake, On representations and compactifications of symmetric Riemannian spaces, Ann. of Math. 71 (1960), 77-110.

[20] E. M. Stein, Note on the boundary values of holomorphic functions, Ann. of Math. 82 (1965), 351-353 .

[20a] E. B. Vinberg, The theory of convex homogeneous cones, Trudy Mosk. Mat. Obs., 12 (1963), 303-358 (in Russian); English translation in Trans. Moscow Math. Soc. for the year 1963, 340-403 .

[21] N. Weiss, Almost everywhere convergence of Poisson intgrals on tube domains over cones, to appear in Trans. Amer. Math. Soc.

[22] J. A. Wolf, Spaces of constant curvature, Mc Graw-Hill, 1967 .

[23] J. A. Wolf, On the classification of Hermitian symmetric spaces, J. of Math. and Mechanics 13 (1964, 489-496).

[24] J. A. Wolf and A. Koranyi, Generalized Cayley transformations of bounded symmetric domains, Amer. J. Math. 87(1965), 899-939.

[25] A. Zygmund, Trigonometric series, Cambridge University Press, 1959 .

CENTRO INTERNAZIONALE MATEMATICO ESTIVO

(C.I.M.E.)

J.L. KOSZUL

" FORMES HARMONIQUES VECTORIELLES SUR LES ESPACES LOCALEMENT SYMETRIQUES"

Sommaire

n. 1 Préliminaires. 199
n. 2 Espaces riemanniens symétriques et espaces fibrés vectoriels associés. 204
n. 3 Seconde structure de fibré associé et connexion linéaire symétrique. 210
n. 4 Formes harmoniques et Laplaciens. 212
n. 5 Décomposition du Laplacien 217
n. 6 Isomorphisme $\mathcal{A}^p(M, E) \longrightarrow \mathcal{A}^p_k(\mathfrak{m}, \mathfrak{v})$ et Théorème de nullité 222
n. 7 Nombres de Betti de M. 229
n. 8 Le cas des domaines bornés symétriques. 233
n. 9 Décomposition du Laplacien dans le cas des domaines bornés symétriques. 238
n. 10 Sous-fibrés holomorphes de E. 245
n. 11 Noyau de Δ_a 250
n. 12 Théorèmes de nullité. 256

Corso tenuto ad Urbino dal 5 al 13 luglio 1967

FORMES HARMONIQUES VECTORIELLES SUR LES ESPACES LOCALEMENT SYMETRIQUES

par

J.L. Koszul

(University of Grenoble)

n. 1 Préliminaires.

Soit M une variété holomorphe compacte admettant pour espace de revêtement universel un domaine borné homogène Ω de C^N, par exemple une surface de Riemann compacte de genre > 1. Soit A le groupe des automorphismes holomorphes de Ω. Puisque Ω est supposé homogène, le groupe A opère transitivement dans Ω. Muni de la topologie "compacts-ouverts", A est un groupe de Lie ; le groupe des automorphismes du revêtement $\Omega \longrightarrow M$ est un sous-groupe discret Γ de A. On sait d'autre part que A opère proprement dans Ω. Choisissons un point $z^o \in \Omega$. Son stabilisateur est un sous-groupe compact K de A. L'application canonique de $\Gamma\backslash A$ sur $\Gamma\backslash A/K$ est donc propre. Or Ω s'identifie à A/K et $\Gamma\backslash A/K$ s'identifie à M. Ceci montre que $\Gamma\backslash A$ est compact, autrement dit que Γ est sous-groupe <u>uniforme</u> de A. On retrouve les mêmes circonstances dans la composante connexe neutre A^o de A. Celle ci opère encore transitivement dans Ω et $\Gamma^o = \Gamma \cap A^o$ est un sous-groupe discret uniforme de A^o. Or un groupe de Lie connexe contenant un sous-groupe discret uniforme est unimodulaire. D'autre part, on sait qu un domaine borné admettant un groupe unimodulaire transitif de transformations holomorphes est un domaine borné symétrique (Théorème de Hano) : tout point de Ω est donc point fixe isolé (et unique) d'un automorphisme involutif holomorphe de Ω.

Montrons maintenant comment, lorsque l'on part d'un domaine borné symétrique Ω, on construit des variétés holomorphes compactes admettant Ω pour revêtement universel. Soit A^o la composante connexe neutre du groupe des automorphismes holomorphes de Ω. Puisque Ω est symétrique, A^o est un groupe de Lie semi-simple réel de centre réduit à l'élément neutre. D'après Borel et Harisch-Chandra, un tel groupe contient toujours des sous-groupes discrets uniformes (dans le cas général, ils s'obtiennent par une construction de nature arithmétique). Soit Γ un sous-groupe discret uniforme de A^o. Puisque A^o opere proprement dans Ω, il en va de même de Γ. Pour tout point $z \in \Omega$, le stabilisateur de z dans Γ est donc un sous-groupe fini de Γ. En général Γ n'opère pas librement dans Ω, c'est à dire qu'il existe des points de Ω dont le stabilisateur dans Γ n'est pas réduit à l'élément neutre. Pour cette raison $\Gamma \backslash \Omega$ ne sera pas en général une variété. Cependant Γ est un groupe linéaire (car A^o est un groupe linéaire) et du fait que c'est un sous-groupe discret uniforme, on déduit qu'il peut être engendré par un nombre fini d'éléments. Or tout groupe linéaire engendré par un nombre fini d'éléments contient un sous-groupe d'indice fini qui est sans torsion, c'est à dire qui ne contient pas d'autre élement d'ordre fini que l'élement neutre (Théor. de Selberg, [23]) Soit Γ' un sous-groupe d'indice fini sans torsion dans Γ'. Il est clair que Γ' opère librement dans Ω puisqu'il opère proprement et qu'il est sans torsion. D'autre part, Γ' étant d'indice fini dans Γ, Γ' est encore un sous-groupe uniforme de A^o. Puisque Γ' opère proprement et librement dans Ω, il existe sur $\Gamma' \backslash \Omega$ une structure de variété holomorphe (et une seule) telle que l'application canonique $\Omega \longrightarrow \Gamma' \backslash \Omega$ soit un étalement holomorphe. Cette variété $\Gamma' \backslash \Omega$ admet ainsi Ω pour espace de revetement universel, car tout domaine borné homogène est homéomorphe à une boule ouverte. Elle est d'autre part compacte car A^o étant transitif sur Ω, $\Gamma' \backslash \Omega$ s'indentifie à un quotient de $\Gamma' \backslash A^o$.

Soient Ω un domaine borné symétrique, M une variété holomorphe compacte admettant un revetement universel $\Omega \longrightarrow M$ et soit Γ le groupe des automorphismes de ce revêtement. La cohomologie de M a été étudiée dans différents cas et avec des buts variés. Il s'agit soit de la cohomologie à coefficients constants, soit de cohomologie à coefficients tordus, soit de cohomologie à coefficients dans le faisceau des sections holomorphes d'un espace fibré vectoriel holomorphe de base M. Les résultats obtenus sont des résultats ne faisant intervenir aucune hypothèse sur Γ. Par contre ils mettent en jeu la nature des facteurs irréductibles figurant dans une décomposition de Ω en produit de domaines bornés symétriques irréductibles. Indiquons deux énoncés particulièrement importants.

Théorème . (Matsushima). Si Ω est un domaine borné symétrique ne comportant pas de facteur irréductible isomorphe à une boule unité dans un espace C^n, le premier nombre de Betti de M est nul.

Puisque Γ est engendré par un nombre fini d'éléments, le sous-groupe des commutateurs de Γ est donc d'indice fini dans Γ lorque l'hypothèse du Théorème est vérifiée.

Théorème. (Calabi, Vesentini, Borel). Soit Θ le faisceau des champs de vecteurs holomorphes sur les ouverts de M . Si Ω est un domaine borné symétrique sans facteur irréductible isomorphe au disque unité de C , alors $H^1(M, \Theta) = (0)$.

D'après un Théorème de Froehlicher et Nijenhuis ([8]) , il en résulte que, dans les hypothèses du Théorème, la structure holomorphe de M n'admet que des déformations continues triviales, contrairement à ce qui se passe pour les surfaces de Riemann de genre > 2 .

Ces deux théorèmes se démontrent par des techniques de formes harmoniques. Des procédés très voisins on été utilisés dans un contexte réel pour étudier les deformations des sous-groupes discrets. Soient G un groupe de Lie réel connexe et Γ un sous-groupe discret uniforme de G. On sait que Γ est engendré par un nombre fini d'éléments. Une famille $w_1, w_2, \ldots, w_p$ de générateurs défini t une application injective de l'ensemble F des homomorphismes de Γ dans G dans l'espace G^p : à tout f $\in$ F est associé $(f(w_1), f(w_2), \ldots f(w_p)) \in G^p$. La topologie la moins fine sur F rendant cette injection continue est indépendante du choix des générateurs w_i. Le groupe G opère dans F par composition avec les automorphismes intérieurs. Si f^o désigne l'injection identique de Γ dans G, on dit que les déformations de Γ dans G sont triviales lorsque la trajectoire Gf^o de f^o est un voisinage de f^o dans F. Soit $\mathfrak{g}$ l'algèbre de Lie de G. Par restriction, la représentation adjointe de G définit une représentation linéaire de Γ dans $\mathfrak{g}$. On voit facilement que la trivialité des déformations de Γ dans G est en rapport avec l'espace de cohomologie $H^1(\Gamma, ad)$ (cohomologie de Γ à valeurs dans le module de la représentation adjointe). Soit en effet f(t) une famille différentiable à un paramètre d'homomorphismes de Γ dans G telle que $f(0) = f^o$. Pour tout $s \in \Gamma$, $s(t) = f(t)s$ est une courbe différentiable dans G et s'(0) est de la forme a(s) s où a est une application de Γ dans $\mathfrak{g}$. En écrivant à l'ordre 1 que f(t) (ss') = (f(t)s(f(t)s') quels que soient $s, s' \in \Gamma$, on voit que $a(ss') = a(s) + sa(s')s^{-1}$, autrement dit a est un 1-cocycle sur Γ à valeurs dans le module de la représentation adjointe. Si u(t) est une courbe différentiable dans G telle que u(0) soit l'élément neutre de G et si $f(t)s = u(t)\,su(t)^{-1}$ quels que soient $s \in \Gamma$ et $t \in R$, alors $a(s) = u'(0) - su'(0)s^{1}$, ce qui signifie que a est le cobord de la 0-co chaine u(0). En fait; on montre que si $H^1(\Gamma, ad) = 0$ les déformations de Γ dans G sont triviales (cf; [24], [25]). Des résultats complets sur $H^1(\Gamma, ad)$ ont été obtenus notamment par E. Calabi et A. Weil dans le cas où G est un groupe semi-simple de centre fini ([5], [24]). Matsushima et Murakami ont d'autre part donné des conditions très générales entraînant la nullité de $H^p(\Gamma, \rho)$ lorsque G est un groupe semi-simple de centre fini et que ρ est la restriction à Γ d'une reprè-

une technique de formes harmoniques présentant beaucoup d'analogies avec celle qui intervient dans la démonstration du Théorème de Calabi et Vesentini. La différence principale vient de ce que ce dernier Théorème concerne un espace de cohomologie à coefficients dans la faisceau des sections holomorphes d'un espace fibré vectoriel holomorphe alors que dans le cas des espaces $H^p(\Gamma, \rho)$ on se ramène en fin de compte à un espace de cohomologie à coefficients tordus sur une variété compacte. Matsushima et Murakami ont mis en évidence les relations existant entre ces deux types de cohomologie et leur travail montre comment finalement des "Théorèmes de nullité" très divers découlent de formules générales donnant les opérateurs Laplaciens ([15] , [16] , [18] , [19]) . Il faut signaler que ces même formules interviennent dans d'autres questions qui concernent les déformations de sous-groupes discrets Γ d'un groupe de Lie G tels que $\Gamma \backslash G$ soit de mesure finie et en particulier les déformations des sous-groupes arithmétiques des groupes algébriques semi-simples (cf. [1] , [21] , [22]) .

n. 2 Espaces riemanniens symétriques et espaces fibrés vectoriels associés.

On désigne par G un groupe de Lie connexe semi-simple de centre fini, par K un sous-groupe compact maximal de G et par Ω l'espace homogène G/K. La dimension de G/K sera notée N. On commencera par un rappel des principales propriétés de G/K. L'espace G/K est difféomorphe à R^N. Le sous-groupe K est connexe et c'est le sous-groupe des éléments de G invariants par un certain automorphisme involutif σ de G. On notera $\mathfrak{g}$ l'algèbre de Lie de G, le crochet étant celui est donné par le crochet des champs de vecteurs invariants à gauche. L'automorphisme σ définit un automorphisme involutif de l'algèbre de Lie $\mathfrak{g}$ qui sera noté σ'. Le noyau de $\sigma' -1$ est la sous-algèbre $\mathfrak{k}$ de $\mathfrak{g}$ correspondant au sous-groupe K. On notera $\mathfrak{m}$ le noyau de $\sigma'+1$; c'est un supplémentaire de $\mathfrak{k}$ dans $\mathfrak{g}$. On a $[\mathfrak{k},\mathfrak{k}] \subset \mathfrak{k}$ $[\mathfrak{k},\mathfrak{m}] \subset \mathfrak{m}$, $[\mathfrak{m},\mathfrak{m}] \subseteq \mathfrak{k}$

Soit B la forme de Killing de $\mathfrak{g}$. Elle est définie positive sur $\mathfrak{m}$, définie négative sur $\mathfrak{k}$ et invariante par σ' ; on a $B(\mathfrak{k}, \mathfrak{m}) = (0)$. On définit un produit scalaire $\langle \, , \, \rangle$ sur $\mathfrak{g}$ en posant

$$\langle a, b \rangle = - B(\sigma' a, b)$$

quels que soient $a, b \in \mathfrak{g}$. Pour tout $a \in \mathfrak{k}$, on a

$$\langle [a, b], b' \rangle + \langle b, [a, b'] \rangle = 0$$

quels que soient $b, b' \in \mathfrak{g}$ et pour tout $c \in \mathfrak{m}$,

$$\langle [c, b], b' \rangle - \langle b, [c, b'] \rangle = 0$$

quels que soient $b, b' \in \mathfrak{g}$.

Soient $(e_i)_{i=1,2,\ldots,N}$ une base orthonormale de $\mathfrak{m}$ et

$(a_k)_{k=1,2,\ldots n-N}$ une base orthonormale de $\mathfrak{k}$ pour ce produit scalaire.

Quels que soient $b, c \in \mathfrak{m}$, on a

$$\sum_i \langle [b, [c, e_i]], e_i \rangle = \sum_i \sum_k \langle [b, a_k], e_i \rangle \langle [c, e_i], a_k \rangle$$

$$\sum_k \langle [b, [c, a_k]], a_k \rangle = \sum_k \sum_i \langle [b, e_i], a_k \rangle \langle [c, a_k], e_i \rangle$$

$$= \sum_{k,} \sum_i \langle [b, a_k], e_i \rangle \langle [c, e_i], a_k \rangle .$$

Par suite $\langle b, c \rangle = B(b, c) = \mathrm{Tr}\ \mathrm{ad}(b)\,\mathrm{ad}(c) = 2\ \mathrm{Tr}_{\mathfrak{m}}\ \mathrm{ad}(b)\mathrm{ad}(c)$ $= 2\ \mathrm{Tr}_{\mathfrak{k}}\ \mathrm{ad}(b)\mathrm{ad}(c)$. On en déduit les relations :

(1) $$\left(\sum_i \mathrm{ad}(e_i)^2 \right) b = \frac{1}{2} b ,$$

(2) $$\left(\sum_k \mathrm{ad}(a_k)^2 \right) b = -\frac{1}{2} b$$

pour tout $b \in \mathfrak{m}$.

Soit q l'application canonique $s \longrightarrow sK$ de G sur $\Omega = G/K$, et soit $z^o = q(K)$. L'application tangente q^T définit une suite exacte $(0) \longrightarrow \mathfrak{k} \longrightarrow \mathfrak{g} \longrightarrow T_{z^o}\Omega \longrightarrow (0)$, où $T_{z^o}\Omega$ désigne l'espace des vecteurs de Ω d'origine z^o. Par suite q^T définit un isomorphisme de $\mathfrak{m}$ sur $T_{z^o}\Omega$. Il existe sur Ω une métrique riemannienne invariante par G et une seule telle que $|q^T b|^2 = \langle b, b \rangle$ pour tout $b \in \mathfrak{m}$. Cette métrique est une métrique riemannienne symétrique. La variété riemannienne Ω est complète. Pour tout $b \in \mathfrak{m}$, $t \longmapsto \exp(tb)z^o$ est une géodésique de Ω. Pour tout vecteur $u \in T_{z^o}\Omega$, $\exp(tb)u$ est un relèvement parallèle de cette géodesique dans

$T\Omega$(*).

Si l'on identifie $\mathfrak{m}$ et $T_{z_o}\Omega$ au moyen de q^T, le tenseur de courbure de la métrique riemannienne symétrique sur Ω est donné par $R(a,b)c = -[[a,b],c]$ quels que soient $a, b, c \in \mathfrak{m}$. Utilisant cette formule et les propriétés de la forme de Killing, on voit que la courbure riemannienne de Ω est ≤ 0 sur toute direction de plan.

Soit Γ un sous-groupe discret de G ; il opère proprement dans Ω. Par conséquent si Γ est sans torsion (cf. n. 1), il opére librement dans Ω. Réciproquement, si Γ opère librement dans Ω, alors Γ est sans torsion. En effet, Ω étant une variété riemannienne complète simplement connexe et de courbure riemannienne ≤ 0, toute isométrie de Ω qui est d'ordre fini laisse fixe au moins un point de Ω.

Dans la suite on supposera que Γ est un sous-groupe discret, sans torsion et uniforme de G. Puisque $\Gamma\backslash G$ est compact, le quotient $M = \Gamma\backslash G/K = \Gamma\backslash\Omega$ est compact. Puisque Γ opère proprement et librement dans Ω, M est une variété differentiable. Il existe sur M une métrique riemannienne et une seule telle que l'application tangente à l'application canonique $\Omega \longrightarrow M$ soit isométrique. Muni de cette métrique, M est une variété riemannienne localement symétrique. C'est une variété orientable car Γ est un sous-groupe de G qui a été supposé connexe.

On peut considerer Ω comme un espace fibré principal différen-

(*) On conviendra d'identifier une variété différentiable M à la section nulle de TM des vecteurs de M. Si G est un groupe de Lie, TG est un groupe de Lie admettant alors G comme sous-groupe. Si G opère différentiablement à gauche dans une variété M, le groupe TG et par suite G opérent à gauche dans TM. Toutes ces lois de composition seront notées multiplicativement.

tiable de base M et de groupe Γ . On désignera dans la suite par E un espace fibré vectoriel différentiable de base M associé à l'espace fibré principal Ω . Feront donc partie de la structure de E une représentation linéaire ρ de Γ dans un espace vectoriel réel V et une application différentiable $p_\Gamma : \Omega \times V \longrightarrow E$ ayant les propriétés suivantes :

(a) $$p_\Gamma(z,v) = p_\Gamma(rz, \rho(r)z)$$

quels que soient $z \in \Omega$, $v \in V$ et $r \in \Gamma$,

(b) pour tout $z \in \Omega$, $v \mapsto p_\Gamma(z,v)$ est un isomorphisme de l'espace vectoriel V sur la fibre de E au point $\Gamma z \in M$.

L'isomorphisme $v \mapsto p_\Gamma(z,v)$ est appelé le repère associé à z .

Il existe dans E une connexion linéaire D et une seule telle que, pour toute courbe différentiable ψ dans Ω , et tout $v \in V$, les vecteurs $p_\Gamma(\psi(t), v)$ soient des vecteurs parallèles le long de la courbe $\Gamma\psi(t)$ ou encore un relèvement parallèle de $\Gamma\psi(t)$ dans E Si f est une section de $\Omega \longrightarrow M$ sur un ouvert $U \subset M$, pour tout $v \in V$, $x \mapsto p_\Gamma(f(x), v)$ est une section de E sur U dont la différentielle covariante est nulle; la coubure de cette connexion linéaire est donc nulle. On notera D_u (resp. D_X) la dérivation covariante par rapport à un vecteur $u \in TM$ (resp. un champ de vecteurs X sur M) .

Soit $\underline{E}$ le faisceau des sections de E dont la différentielle covariante est nulle . D'après le Théorème de Rham (cas de coefficients tordus), les espaces de cohomologie $H^p(M, \underline{E})$ s'obtiennent comme espaces de cohomologie du complexe

$$\mathcal{A}^0(M,E) \xrightarrow{d} \mathcal{A}^1(M,E) \ldots \mathcal{A}^p(M,E) \xrightarrow{d} \mathcal{A}^{p+1}(M,E) \ldots$$

où $\mathcal{A}^p(M,E)$ désigne l'espace des formes différentielles alternées de de-

gré p sur M à valeurs dans E et où d est l'opérateur de differentiation extérieure défini par la connexion D .
Si $\varphi \in \mathcal{A}^{o}(M,E)$, c'est à dire si φ est une section différentiable de E, $d\varphi$ est la différentielle covariante de φ . Si $\omega \in \mathcal{A}^{p}(M,E)$ et si ω est égale à $\sum_{i=1}^{i=r} \omega_i \varphi_i$ où les ω_i sont des p-formes scalaires et les φ_i des sections de E, alors $d\omega = \sum_i (d\omega_i)\varphi_i + (-1)^p \sum_i \omega_i \wedge (d\varphi_i)$. On peut également caractériser $d\omega$ par la condition

$$(d\omega)(X_1, X_2, \ldots X_{p+1}) = \sum_i (-1)^{i+1} D_{X_i}\, \omega(X_1, \ldots \hat{X}_i, \ldots X_{p+1})$$

$$+ \sum_{i<j} (-1)^{i+j}\, \omega([X_i, X_j], \ldots \hat{X}_i, \ldots \hat{X}_j, \ldots X_{p+1}),$$

pour toute suite de p champs de vecteurs X_i sur M .
Théorème 1(S. Eilenberg) . Il existe pour tout p un isomorphisme canonique de $H^p(M,\underline{E})$ sur l'espace $H^p(\Gamma, \rho)$ des classes de cohomologie de Γ à coefficients dans l'espace V de la représentation linéaire ρ .

Pour tout p , soit $\mathcal{A}^{p}(\Omega, V)$ l'espace des formes différentielles de degré p sur Ω à valeurs dans V. On définit une représentation linéaire $\hat{\rho}$ de Γ dans $\mathcal{A}^{p}(\Omega, V)$ en posant

$$(\hat{\rho}(r)\,\omega)(w_1, w_2, \ldots w_p) = \rho(r)\,\omega(\bar{r}^{1} w_1, \bar{r}^{1} w_2, \ldots \bar{r}^{1} w_p)$$

pour tout $r \in \Gamma$, toute forme $\omega \in \mathcal{A}^{p}(\Omega, V)$ et toute suite de vecteurs $w_1, w_2, \ldots w_p \in T\Omega$ ayant même origine. Par un procédé analogue à celui qui sera explicité au n. 6 , on peut identifier l'espace $\mathcal{A}^{p}(M,E)$ au sous-espace $\mathcal{A}^{p}_{\Gamma}(\Omega, V)$ des éléments de $\mathcal{A}^{p}(\Omega, V)$ invariants par Γ .
Après cette identification, la différentiation extérieure définie plus haut sur $\mathcal{A}^{p}(M,E)$ devient la restriction à $\mathcal{A}^{p}_{\Gamma}(\Omega, V)$ de l'opérateur de différentiation extérieure des p-formes à valeurs dans V sur Ω . Pour tout couple d'entiers (p, q) , soit $C^{p,q}$ l'espace des p-cochaînes sur Γ

à valeurs dans le Γ-module $\mathcal{Q}^q(\Omega, V)$. On a deux opérateurs de carré nul

$$d_\Omega : C^{p,q} \longrightarrow C^{p,q+1}$$
$$d_\Gamma : C^{p,q} \longrightarrow C^{p+1,q}$$

provenant respectivement de la différentiation extérieure $d: \mathcal{Q}^q(\Omega, V) \to \mathcal{Q}^{q+1}(\Omega, V)$ et de l'opérateur bord du complexe des cochaînes sur Γ à valeurs dans le Γ-module $\mathcal{Q}^q(\Omega, V)$. Comme $d: \mathcal{Q}^q(\Omega, V) \to \mathcal{Q}^{q+1}(\Omega, V)$ est un homomorphisme de Γ-modules, on a $d_\Omega d_\Gamma = d_\Gamma d_\Omega$. On définit donc un complexe

$$D^o \xrightarrow{d} D^1 \dots D^r \xrightarrow{d} D^{r+1} \dots$$

en posant $D^r = \bigoplus_{p+q=r} C^{p,q}$ et $d = d_\Gamma + (-1)^p d_\Omega$ sur $C^{p,q}$.

Les suites exactes

$$(0) \longrightarrow \mathcal{Q}^q_\Gamma(\Omega, V) \longrightarrow C^{o,p} \xrightarrow{d_\Gamma} C^{1,q} \qquad (q=o, 1, \dots)$$

$$(0) \longrightarrow C^p(\Gamma, V) \longrightarrow C^{p,o} \xrightarrow{d_\Omega} C^{p,1} \qquad (p, =o, 1, \dots)$$

(où $C^p(\Gamma, V)$ désigne l'éspace des p-cochaînes sur Γ à valeurs dans V) montrent respectivement qu'il existe des homomorphismes injectifs canoniques $H^q(M, \underline{E}) \longrightarrow H^q(D^*)$ et $H^p(\Gamma, \rho) \longrightarrow H^p(D^*)$ quels que soient p et q. Du fait que Γ opère proprement et librement dans Ω résulte que les premiers sont surjectifs. Du fait que Ω est un espace contractile résulte que les seconds sont surjectifs.

On notera que le groupe transitif G n'intervient pas dans la démonstration. Le résultat est vrai toutes les fois que Γ est un groupe discret opérant différentiablement et proprement dans une variété contractile Ω

n. 3 Seconde structure de fibré associé et connexion linéaire symétrique.

On conserve les hypothèses et les notations du n. 2 , mais on suppose que la représentation ρ de Γ dans V est définie comme la restriction à Γ d'une représentation linéaire de G dans V que l'on note également ρ .

Puisque Γ opère librement dans $\Omega = G/K$, $\Gamma \backslash G$ est un espace fibré principal de base M et de groupe K (opérant à droite). On va voir que $\rho : G \longrightarrow Gl(V)$ et p_Γ définissent dans l'espace fibré vectoriel E une structure d'espace fibré vectoriel associé à l'espace fibré principal $\Gamma \backslash G$. On définit une application différentiable p_K de $(\Gamma\backslash G) \times V$ dans E en posant

(1) $$p_K(\Gamma s, v) = p_\Gamma (sK, \rho(s)v)$$

quels que soient $s \in G$ et $v \in V$. En effet, $p_\Gamma(sK, \rho(s)v)$ ne dépend que de la classe Γs de s. Pour tout $r \in K$, on a

(2) $$p_K(\Gamma sr^{-1}, \rho(r)v) = p_\Gamma (sK, \rho(sr^{-1})\rho(r)v) = p_K(\Gamma s, v)$$

quels que soient $s \in G$ et $v \in V$. D'autre part, quel que soit $\Gamma s \in \Gamma\backslash G$, $v \longmapsto p_K(\Gamma s, v)$ est un isomorphisme de V sur la fibre de E au point ΓsK de M . L'application p_K et la restriction de ρ à K définissent donc dans E une structure d'espace vectoriel associé à l'espace fibré principal $\Gamma\backslash G$.

En tout point $\xi \in \Gamma\backslash G$, l'espace $T_\xi(\Gamma\backslash G)$ des vecteurs d'origine ξ est somme directe des sous-espace $\xi\mathfrak{k}$ et $\xi\mathfrak{m}$. Cette décomposition est stable par les opérations de K . L'espace $\xi\mathfrak{k}$ est l'espace des vecteurs d'origine ξ tangents à la trajectoire ξK de ξ . Il existe dans E une connexion linéaire D^s et une seule qui vérifie la condition suivante : pour tout $v \in V$ et toute courbe différentiable ψ dans $\Gamma\backslash G$ telle que $\psi'(t) \in \psi(t)\mathfrak{m}$, $t \longmapsto p_K(\psi(t), v)$ est un relèvement parallèle de la courbe $\psi(t)K$ dans le fibré E. Cette connexion D^s sera appelée la <u>connexion symétrique</u>. On notera D^s_u (resp. D^s_X) la dérivation covariante par rapport à un vecteur $u \in TM$ (resp. un champ de vecteurs X sur M) définie par la connexion symétrique. On définit comme au n. 2 l'opérateur de différentiation extérieure d_s relatif à D^s. On sait que si R désigne la forme de courbure de D^s, pour toute forme $\omega \in \mathfrak{a}^p(M, E)$, on a $(d_s)^2 \omega = R \wedge \omega$. La forme de courbure R n'est pas nulle et par suite $(d_s)^2 \neq 0$.

Puisque K est compact, il existe sur V des produits scalaires invariants par $\rho(r)$ pour tout $r \in K$. Dans la suite, on supposera choisi un tel produit scalaire que l'on notera $\langle \, , \, \rangle$. Il existe alors dans chaque fibre de E un produit scalaire et un seul tel que, pour tout $\xi \in \Gamma\backslash G$, le repère $v \longmapsto p_K(\xi, v)$ soit une isométrie de V sur la fibre au dessus de ξK. Ce produit scalaire dans les fibres de E s'interprète comme un homomorphisme différentiable μ de $E \underset{M}{\otimes} E$ dans $M \times R$ considéré comme espace fibré vectoriel de base M. On voit facilement qu'il est invariant par transport parallèle (relativement à D^s).

N. 4 Formes harmoniques et Laplaciens.

Dans ce N. on désignera par M une variété riemannienne connexe orientée de dimension N. On notera v la forme volume positive sur M, c'est à dire la forme différentielle de degré N telle que $v(e_1, e_2, \dots e_N) = 1$ lorsque $e_1, e_2, \dots e_n$ est une suite orthonormale positive de vecteurs en un point de M. On désignera par F l'algèbre des fonctions différentiables sur M.

Soit E un espace fibré vectoriel différentiable de base M . Pour tout entier p, on notera $\mathcal{Q}^p(M, E)$ le F-module des formes différentielles alternées de degré p sur M à valeurs dans E . Pour tout champ de vecteurs X sur M, on notera γ_X la forme différentielle scalaire de degré 1 que la métrique riemannienne associe à X . On notera (X) (resp. $\varepsilon(\gamma_X)$) la multiplication intérieure à gauche par X (resp. La multiplication extérieure à gauche par γ_X). Pour tout p il existe un homomorphisme canonique $\omega \longmapsto *\omega$ du F-module $\mathcal{Q}^p(M, E)$ dans le F-module $\mathcal{Q}^{N-p}(M, E)$. La famille des ces homomorphismes est caractérisée par les propriétés suivantes :

(a) si $\varphi \in \mathcal{Q}^o(M, E)$, alors $*\varphi = \varphi v$,

(b) pour tout champ de vecteurs X sur M et toute forme $\omega \in \mathcal{Q}^p(M, E)$

$$*\, \varepsilon(\gamma_X)\, \omega = (-1)^p\, \iota(X) * \omega \quad .$$

On vérifie que

$$**\omega = (-1)^{p(N-p)}\, \omega$$

pour toute forme ω de degré p. Les applications $\omega \longmapsto *\omega$ sont donc bijectives.

Supposons maintenant défini un produit scalaires $\mu : E \underset{M}{\otimes} E \longrightarrow M \times R$ dans les fibres de E .

Soient ω et ω' des formes différentielles sur M à valeurs dans E de degrés respectifs p et q. Leur produit extérieur est une (p+q)-forme $\omega \wedge \omega'$ à valeurs dans $E \underset{M}{\otimes} E$. En composant cette forme avec μ, on obtient une (p+q)-forme scalaire sur M qui sera notée $\langle \omega \wedge \omega' \rangle$. Si $\omega, \omega' \in \mathcal{Q}^p(M,E)$, alors $\langle \omega \wedge (* \omega') \rangle$ est de degré N ; c'est donc le produit de v par une fonction $(\omega, \omega') \in F$. On a ainsi obtenu une forme bilinéaire sur le F-module $\mathcal{Q}^p(M,E)$ à valeurs dans F. On vérifie facilement que cette forme est symétrique. Si $p = 0$, elle associe à déux sections la fonction obtenue en prenant dans chaque fibre de E le produit scalaire des valeurs. Soit $(X_i)_{i=1,2...N}$ une base orthonormale de champs de vecteurs sur un ouvert U de M. Si $\omega, \omega' \in \mathcal{Q}^p(M,E)$ et si $p > 0$, la restriction de (ω, ω') à U est donnéepar la formule

$$(\omega, \omega') = \sum_{\tau} (\omega \; (X_{\tau 1}, X_{\tau 2}, \ldots X_{\tau p}), \; \omega' (X_{\tau 1}, X_{\tau 2}, \ldots X_{\tau p}))$$

où τ parcourt l'ensemble des applications strictement croissantes de $[1, p]$ dans $[1, N]$. Il en résulte que $(\omega, \omega) > 0$ en tout point où $\omega \neq 0$.

On suppose maintenant donnée dans E une connexion linéaire D^s telle que le produit scalaire dans les fibres soit invariant par transport parallèle. Cette condition est équivalente à la condition ;

$$(1) \qquad X.(\varphi, \psi) = (D^s_X \varphi, \psi) + (\varphi, D^s_X \varphi)$$

pour tout champ de vecteurs X sur M et tout couple de sections $\varphi, \psi \in \mathcal{Q}^o(M,E)$. Cette connexion définit pour tout p un opérateur de différentiation extérieure $d_s : \mathcal{Q}^p(M,E) \longrightarrow \mathcal{Q}^{p+1}(M,E)$.
On définit, pour tout p, un opérateur différentiel $\partial_s : \mathcal{Q}^p(M,E) \longrightarrow \mathcal{Q}^{p-1}(M,E)$ en posant

$$\partial_s \omega = (-1)^{p(N-p+1)} * d_s * \omega$$

pour toute forme $\omega \in \mathcal{A}^p(M,E)$. Il résulte de (1) que

(2) $$d\langle \omega \wedge * \omega' \rangle = (d_s \omega , \omega')v - (\omega , \partial_s \omega')v$$

quelles que soient ω de degré p et ω' de degré p+1 .

Pour expliciter, localement, l'opérateur ∂_s , on introduira la connexion linéaire ∇ dans TM qui est définie par la métrique riemannienne sur M. Cette connexion ∇ et la connexion D^s définissent une connexion linéaire dans l'espace fibré vectoriel $\Lambda^p(TM)^* \underset{M}{\otimes} E$. Les sections différentiables de ce fibré étant les éléments de $\mathcal{A}^p(M,E)$, on définit ainsi la dérivation covariante dans $\mathcal{A}^p(M,E)$ par rapport à un champ de vecteurs X sur M . Cette dérivation covariante sera notée D^s_X (pour p=0 , on retrouve la dérivation covariante des sections de E relative à la connexion D^s). Quels que soient les champs de vecteurs $X, Y_1, \ldots Y_p$ sur M, on a

$$(D^s_X \omega)(Y_1, Y_2, \ldots Y_p) = D^s_X(\omega(Y_1, Y_2, \ldots Y_p)) - \sum_i \omega(Y_1, \ldots \nabla_X Y_i, \ldots Y_p)$$

pour toute forme $\omega \in \mathcal{A}^p(M,E)$. Cela étant , si $(X_i)_{i=1,2,..N}$ est une base orthonormale de champs de vecteurs sur un ouvert U de M et si γ_i est la base duale de 1-formes, on a sur U :

$$d_s = \sum_i \epsilon(\gamma_i) D^s_{X_i}$$

$$\partial_s = -\sum_i L(X_i) D^s_{X_i} \quad .$$

Soit maintenant D une seconde connexion linéaire dans E dont on supposera la <u>courbure nulle.</u> Soit A la forme différentielle de degré 1 sur M à valeurs dans l'espace fibré End(E) definie par $A(u) = D_u - D^s_u$ pour tout $u \in TM$. On notera A^* la forme transposée : pour tout champ de vecteurs X sur M et tout couple de sections $\varphi, \psi \in \mathcal{Q}^0(M,E)$, on a $(A^*(X)^*\varphi, \psi) = (\varphi, A(X)\psi)$. On vérifie facilement que l'opérateur $d_a = d - d_s$ transforme une forme $\omega \in \mathcal{Q}^p(M,E)$ en la forme composée de $A \wedge \omega$ et de l'homomorphisme canonique $\mathrm{End}(E) \otimes_M E \longrightarrow E$. Il existe un opérateur ∂_a de degré -1 et un seul tel que

(3) $$(d_a \omega, \omega') = (\omega, \partial_a \omega')$$

quelles que soient $\omega \in \mathcal{Q}^p(M,E)$ et $\omega' \in \mathcal{Q}^{p+1}(M,E)$. Localement, avec les mêmes notations que plus haut , on a

$$d_a = \sum_i \varepsilon(\gamma_i) A(X_i) ,$$
$$\partial_a = \sum_i L(X_i)\, A^*(X_i) .$$

On posera $\partial = \partial_s + \partial_a$, $\Delta = d\partial + \partial d$, $\Delta_s = d_s \partial_s + \partial_s d_s$ et $\Delta = d_a \partial_a + \partial_a d_a$. Les opérateurs Δ, Δ_s et Δ_a sont appelés des <u>opérateurs Laplaciens</u>. Les formes annulées par Δ sont appelées des <u>formes harmoniques</u>.

Supposons que M soit compacte. On definit alors pour tout p un produit scalaire $\langle\, ,\rangle$ sur $\mathcal{Q}^p(M,E)$ en posant

$$\langle \omega, \omega' \rangle = \int_M (\omega, \omega')v = \int_M \langle \omega \wedge (\omega') \rangle .$$

Il résulte des relations (2) et (3) que

$$\langle d_s \omega, \omega' \rangle = \langle \omega, \partial_s \omega' \rangle ,$$
$$\langle d_a \omega, \omega' \rangle = \langle \omega, \partial_a \omega' \rangle ,$$

ce qui donne

$$\langle d\omega , \omega' \rangle = \langle \omega , \partial \omega' \rangle$$

quelles que soient les formes ω, ω'. Par conséquent

$$\langle \Delta \omega, \omega \rangle = \langle d\omega, d\omega \rangle + \langle \partial \omega, \partial \omega \rangle ,$$

ce qui montre que les conditions suivantes sont équivalentes :

(a) $\langle \Delta \omega, \omega \rangle = 0$,

(b) $\Delta \omega = 0$,

(c) $d\omega = \partial \omega = 0$.

On a donc une application linéaire injective de l'espace des formes harmoniques de degré p dans le p-ième espace de cohomologie du complexe

$$\mathcal{A}^{0}(M, E) \xrightarrow{d} \mathcal{A}^{1}(M, E) \xrightarrow{d} \mathcal{A}^{2}(M, E) \xrightarrow{d} \dots$$

D'après le Théorème de Hodge Kodaira (à coefficients tordus) cette application est un isomorphisme et ces espaces sont de dimensions finies (cf. [2]) . On obtient donc en définitive un <u>isomorphisme canonique de l'espace des formes harmoniques de degré</u> p <u>sur l'espace</u> $H^p(M, \underline{E})$ où $\underline{E}$ désigne la faisceau des sections de E ayant une différentielle covariante nulle.

On voit de même que $\langle \Delta_s \omega , \omega \rangle = \langle d_s \omega, d_s \omega \rangle + \langle \partial_s \omega, \partial_s \omega \rangle$ et $\langle \Delta_a \omega , \omega \rangle = \langle d_a \omega, d_a \omega \rangle + \langle \partial_a \omega, \partial_a \omega \rangle$, ce qui donne pour les zéros de Δ_s et de Δ_a des caractérisations analogues à celles que l'on vient de voir pour les zéros de Δ .

N°5 Décomposition du Laplacien.

Les hypothèses et les notations sont celles du N° précédent ; M n'est pas supposée compacte.

Lorsque l'on explicite le Laplacien Δ avec les opérateurs d_s, d_a, ∂_s, ∂_a, on obtient

$$\Delta = \Delta_s + \Delta_a + d_s \partial_a + \partial_a d_s + \partial_s d_a + d_a \partial_s .$$

On dira qu'il y a décomposition du Laplacien Δ lorsque $\Delta = \Delta_s + \Delta_a$. Si cette relation est verifiée, puisque $(\Delta_s \omega, \omega) \geqslant 0$ et $(\Delta_a \omega, \omega) \geqslant 0$ pour toute forme ω, on voit que toute forme harmonique appartient au noyau de Δ_a. Or Δ_a est un opérateur différentiel d'ordre 0. Lorsqu'il y a décomposition du Laplacien Δ, on obtiendra ainsi des conditions de nature "algébrique" qui devront être vérifiées par toute forme harmonique.

Lemme 1. *Pour que $\Delta = \Delta_s + \Delta_a$ il faut et il suffit que les conditions suivantes soient vérifiées :*

(a) $A^* = A$

(b) *pour toute géodésique φ dans M*, $[D_{\varphi'(t)}, A(\varphi'(t))] = 0$ *quel que soit* $t \in \mathbb{R}$.

Si φ est une géodésique dans M, pour tout $t \in \mathbb{R}$ on a un endomorphisme $A(\varphi'(t))$ dans la fibre au point $\varphi(t)$. La condition (b) signifie que ces endomorphismes se déduisent les uns des autres par transport parallèle relativement à D le long de la géodésique φ. Puisque $D_u = D^s_u + A(u)$ pour tout $u \in TM$, il revient au même de dire que ces endomorphismes se déduisent les uns des autres par transport parallèle relativement à D^s.

La démonstration du Lemme 1 se fait par un calcul de $d_s \partial_a + \partial_a d_s + d_a \partial_s + \partial_s d_a$ utilisant les expressions locales de ces opérateurs données au n. 4.

On détaillera dans ce qui suit le cas où dim M = 1 qui est plus parlant que le cas général.

Supposons donc que $M = \mathbb{R}$ et que E soit l'espace fibré vectoriel trivial $\mathbb{R} \times \mathbb{R}^{\ell}$. On désigne par P le champ de vecteurs d/dt sur $\mathbb{R}$ et on munit $\mathbb{R}$ de la métrique riemannienne euclidienne qui fait de P un champ de vecteurs unitaires. Le produit scalaire dans les fibres sera defini par une matrice carrée symétrique définie positive d'ordre ℓ fonction différentiable de $t \in \mathbb{R}$ que l'on notera S. Les sections de E (resp. les formes différentielles à valeurs dans E) seront identifiées à des fonctions (resp. des formes) à valeurs dans $\mathbb{R}^{\ell}$. On supposera que la connexion D est telle que $D_P \varphi = P.\varphi$ pour toute section φ de E. On posera $D^s_P \varphi = P.\varphi - A$ où A est une matrice carrée d'ordre ℓ fonction différentiable de t. Si le produit scalaire dans les fibres est invariant par transport parallèle relatif à la connexion D^s, on a, en notant ${}^tX$ la transposée d'une matrice X :

$$P.({}^t\varphi S \psi) = {}^t(P.\varphi - A\varphi)S\psi + \varphi^t \, S(P.\psi - \psi)$$

quelles que soient les fonctions φ et ψ à valeurs dans $\mathbb{R}^{\ell}$. En posant $A^* = S^{-1}({}^tA)S$, la relation précédente s'écrit :

$$S^{-1}(P.S.) + A + A^* = 0 . \tag{1}$$

Pour toute fonction φ sur $\mathbb{R}$ à valeurs dans R^{ℓ}, on a :

$$d_s\varphi = (P.\varphi)dt - (A\varphi)dt \quad , \quad d_a\varphi = (A\varphi)dt \quad ,$$

$$d\varphi = (P.\varphi)dt \quad ,$$

$$\partial_s(\varphi\, dt) = - P.\varphi + A\varphi , \quad \partial_a(\varphi\, dt) = A\varphi ,$$

$$\partial(\varphi\, dt) = - P.\varphi + (A + A^*)\varphi ,$$

$$\Delta_s\varphi = - P^2.\varphi + P.(A\varphi) + A(P.\varphi) - A^2\varphi \quad ;$$

$$\Delta_a\varphi = A^*A\varphi \quad ,$$

$$\Delta\varphi = - P^2.\varphi + (A + A^*)(P.\varphi) .$$

Si $\Delta = \Delta_s + \Delta_a$, on doit avoir

$$(P.A)\varphi + 2A(P.\varphi) - (A+A^*)(P.\varphi) - A^2\varphi + A^*A\varphi = 0$$

pour toute fonction φ sur $\mathbb{R}$ à valeurs dans $\mathbb{R}^\ell$.
Ceci implique $A = A^*$ et $P.A = 0$. Puisque

$$D^s_P A\varphi - A D^s_P \varphi = P.(A\varphi) - A^2\varphi - A(P.\varphi) + A^2 = (P.A)\varphi$$

la condition P.A. = signifie que la famille d'endomorphismes ayant pour matrice A est invariante par transport parallèle relativement à la connexion D^s . Réciproquement, si $A = A^*$ et si P.A. = 0, le calcul précédent montre que $\Delta = \Delta_s + \Delta_a$ sur les formes de degré 0 et on voit facilement qu'elle est alors également vérifiée sur les formes de degré 1 .

On revient maintenant aux notations et aux hypothèses des n. 2 et 3 . On a donc un espace fibré vectoriel E de de base M = $= \Gamma \backslash G/K$ associé à G/K , la représentation linéaire ρ de Γ dans V étant la restriction d'une representation linéaire ρ de G dans V . On a défini deux connexions linéaires dans E : la connexion D dont la courbure est nulle et la connexion symétrique D^s . On suppose choisi un produit scalaire $\langle , \rangle$ sur V, invariant par $\rho(r)$ pour tout $r \in K$. En désignant encore par ρ la représentation linéaire de l'algèbre de Lie $\mathfrak{g}$ définie par ρ on a

(2) $$\langle \rho(a) v , v' \rangle + \langle v, \rho(a)v' \rangle = 0$$

quels que soient $v, v' \in V$ et $a \in \mathfrak{k}$. On a vu qu'il existe un produit scalaire et un seul dans les fibres de E tel que pour tout $\xi \in \Gamma \backslash G$, le repère $v \mapsto p_K(\xi, v)$ soit isométrique . Utilisant ce produit sclaire, la métrique riemannienne symétrique sur M et les connexions linéaires dans E on définit comme au n. 4 les Laplaciens Δ , Δ_s et Δ_a.

Lemme 2. Pour que $\Delta = \Delta_s + \Delta_a$, il faut et il suffit que

(3) $$\langle \rho(b)\, v, v' \rangle - \langle v, \rho(b)\, v' \rangle = 0$$

quels que soient $v, v' \in V$ et $b \in \mathfrak{m}$.

Soit en effet φ une géodésique de M. Elle est de la forme $\varphi(t) = \Gamma\, s\underline{\exp}(tb)\, K$ où $s \in G$ et $b \in \mathfrak{m}$. Soit $r_1(t)$ la famille de repères le long de φ définie par $r_1(t)\, v = p_\Gamma(s\underline{\exp}(tb)K, v)$ pour tout $t \in \mathbb{R}$ et tout $v \in V$. Ces repères sont parallèles pour la connexion D car $s\underline{\exp}(tb)K$ est un relèvement de φ dans G/K. Les repères $r(t)$ définis par $r(t)v = r_1(t)\, \rho(s)v = p_\Gamma(s\underline{\exp}(tb)K, \rho(s)v)$ sont donc également des repères parallèles pour la connexion D. La courbe $\psi(t) = \Gamma\, s\underline{\exp}(tb)$ dans $\Gamma\backslash G$ vérifie la condition $\psi'(t) \in \psi(t)\, \mathfrak{m}$ pour tout $t \in R$. Par conséquent, les repères $r_s(t)$ définis par $r_s(t)v = p_K(s\underline{\exp}(tb), v)$ sont des repères parallèles le long de φ pour la connexion symétrique D^s. On a $r(t)\, \rho(\underline{\exp}(tb)) = r_s(t)$ pour tout t. Il en résulte que

$$(D_{\varphi'(t_0)} - D^s_{\varphi'(t_0)})\, r(t)v = - D^s_{\varphi'(t_0)}\, r(t)v$$

$$= - D^s_{\varphi'(t_0)}\, r_s(t)\, \rho(\exp(-tb))v = r_s(t_0)\, \rho(\exp(-t_0 b))\, \rho(b)\, v$$

$$= r(t_0)\, \rho(b)\, v$$

pour tout $t_0 \in \mathbb{R}$. La forme A est donc telle que $A(\varphi'(t))\, r(t) = r(t)\, \rho(b)$ pour tout t. Ceci montre que les $A(\varphi'(t))$ se déduisent les uns des autres par transport parallèle le long de la géodésique φ. On a de même $A(\varphi'(t))\, r_s(t) = r_s(t)\, \rho(b)$ pour tout t. Puisque les repères $r_s(t)$ sont des isométries de V sur les fibres de E, on voit que la relation $A(\varphi'(t)) = A^*(\varphi'(t))$ pour tout $t \in \mathbb{R}$ est équivalente à $\langle \rho(b)v, v' \rangle = \langle v, \rho(b)\, v' \rangle$ quels que soient $v, v' \in V$. Compte tenu du Lemme 1, ceci achève la démonstration.

Pour toute représentation ρ de G dans V, il existe un produit scalaire $\langle , \rangle$ sur V qui vérifie les conditions (2) et (3). On le voit en construisant la représentation linéaire ρ^c de $\mathfrak{g}^c = \mathfrak{g} + \sqrt{-1}\,\mathfrak{g}$ dans $V^c = V + \sqrt{-1}\,V$ déduite de ρ par extension des scalaires. La sous-algèbre $\mathfrak{k} + \sqrt{-1}\,\mathfrak{m}$ est une sous-algèbre semi-simple compacte de $\mathfrak{g}^c$. Il existe donc une forme hermitienne définie positive h sur V^c telle que $h(\rho(x)w, w') + h(w, \rho(x)w') = 0$ quels que soient $x \in \mathfrak{g}^c$ et $w, w' \in V^c$. La restriction à V de la partie réelle de h est un produit scalaire sur V qui vérifie les conditions (2) et (3).

n. 6. Isomorphismes $\mathfrak{a}^p(M,E) \to \mathfrak{a}^p_k(\mathfrak{m},\mathcal{V})$ et Théorème de nullité.

Pour étudier les formes sur M à valeurs dans E et les opérateurs Laplaciens sur ces formes, il est commode de les identifier à des formes alternées sur $\mathfrak{m}$ à valeurs dans l'espace $\mathcal{V}$ constitué par les applications différentiables de $\Gamma\backslash G$ dans V.

Pour tout vecteur $w \in T(\Gamma\backslash G)$ notons wK l'image de w par l'application tangente à l'application canonique de $\Gamma\backslash G$ sur $M = \Gamma\backslash G/K$. Pour tout point $\xi = \Gamma s$, $b \to (\xi b)K$ est un isomorphisme de $\mathfrak{m}$ sur l'espace des vecteurs d'origine ΓsK dans TM. Si $\omega \in \mathfrak{a}^p(M,E)$, l'application $(b_1, b_2, \ldots b_p) \to ((\xi b_1)K, (\xi b_2)K, \ldots (\xi b_p)K)$ est une forme alternée de degré p sur $\mathfrak{m}$ à valeurs dans la fibre de E qui est au dessus du point ξK. On définit donc pour tout $\xi \in \Gamma\backslash G$ une forme alternée $\bar\omega_\xi$ de degré p sur $\mathfrak{m}$ à valeurs dans V en posant

$$p_K(\xi, \omega_\xi(b_1, b_2, \ldots b_p)) = \omega((\xi b_1)K, (\xi b_2)K, \ldots (\xi b_p)K)$$

pour toute suite $b_1, b_2, \ldots b_p \in \mathfrak{m}$. Quelle que soit la suite $b_1, b_2, \ldots b_p \in \mathfrak{m}$ l'application $\xi \mapsto \bar\omega_\xi(b_1, b_2, \ldots b_p)$ est une application différentiable de $\Gamma\backslash G$ dans V ; on la notera $\bar\omega(b_1, b_2, \ldots b_p)$. Soit $\mathfrak{a}^p(\mathfrak{m}, \mathcal{V})$ l'espace des formes alternées de degré p sur $\mathfrak{m}$ à valeurs dans $\mathcal{V}$. On a obtenu une application linéaire canonique injective $\omega \mapsto \bar\omega$ de $\mathfrak{a}^p(M,E)$ dans $\mathfrak{a}^p(\mathfrak{m}, \mathcal{V})$. Compte tenu des propriétés de p_K on voit que

$$(1) \qquad \bar\omega_{\xi r}(r^{-1} b_1 r, r^{-1} b_2 r, \ldots r^{-1} b_p r) = \rho(r^{-1})\, \bar\omega_\xi(b_1, b_2, \ldots b_p)$$

quels que soient $\xi \in \Gamma\backslash G$, $r \in K$ et $b_1, b_2, \ldots b_p \in \mathfrak{m}$

On vérifie que l'image de l'application canonique $\mathfrak{a}^p(M,E) \to \mathfrak{a}^p(\mathfrak{m}\ \mathcal{V})$ est le sous-espace $\mathfrak{a}^p_k(\mathfrak{m}, \mathcal{V})$ constitué par les formes $\bar\omega$ qui satisfont à la condition (1). On identifiera dans la suite les éléments de $\mathfrak{a}^p(M, E)$ et leurs images dans $\mathfrak{a}^p(\mathfrak{m}, \mathfrak{V})$.

Pour tout $x \in \mathfrak{g}$ on note L_x le champ de vecteurs sur $\Gamma\backslash G$ dont la valeur en ξ est ξx. C'est la projection sur $\Gamma\backslash G$ du champ de vecteurs invariant à gauche sur G ayant x pour valeur en l'elément neutre de G. On a donc $L_{[x,y]} = [L_x, L_y]$ quels que soient $x, y \in \mathfrak{g}$.

On définit dans $\mathfrak{a}^p(\mathfrak{m}, \mathfrak{V})$ deux structures de $\mathfrak{g}$ modules.

La première fait passer de $(x, \bar{\omega}) \in \mathfrak{g} \times \mathfrak{a}^p(\mathfrak{m}, \mathfrak{V})$ à la forme $\rho(x)\bar{\omega}$ définie par

(2) $$(\rho(x)\bar{\omega})_\xi (b_1, b_2, \ldots b_p) = \rho(x)(\omega_\xi(b_1, b_2, \ldots b_p))$$

quels que soient $\xi \in \Gamma\backslash G$ et $b_1, b_2, \ldots b_p \in \mathfrak{m}$. La seconde fait passer de $(x, \bar{\omega})$ à $L_x\bar{\omega}$ définie par

(3) $$(L_x\bar{\omega})(b_1, b_2, \ldots b_p) = L_x.\bar{\omega}(b_1, b_2, \ldots b_p)$$

où $L_x.\bar{\omega}(b_1, b_2, \ldots b_p)$ désigne la dérivée de Lie de la fonction $\bar{\omega}(b_1, b_2, \ldots b_p)$ par rapport au champ de vecteurs L_x.

Compte tenu de l'inclusion $[\mathfrak{k}, \mathfrak{m}] \subset \mathfrak{m}$, on définit enfin dans $\mathfrak{a}^p(\mathfrak{m}. \mathfrak{V})$ une structure de $\mathfrak{k}$-module qui fait passer de $(a, \bar{\omega}) \in \mathfrak{k} \times \mathfrak{a}^p(\mathfrak{m}, \mathfrak{V})$ à la forme $\theta(a)\bar{\omega}$ définie par

(4) $$(\theta(a)\bar{\omega})(b_1, b_2, \ldots b_p) = -\sum_i \bar{\omega}(b_1, \ldots [a, b_i], \ldots b_p)$$

quels que soient $b_1, b_2, \ldots b_p \in \mathfrak{m}$.

Il est clair que, quels que soient $x, y \in \mathfrak{g}$ et $a \in \mathfrak{k}$, les endomorphismes $\rho(x)$, L_y et $\theta(a)$ de $\mathfrak{a}^p(\mathfrak{m}\ \mathfrak{V})$ commutent deux à deux. Par suite, les endomorphismes $\rho(a) + L_a + \theta(a)$ où $a \in \mathfrak{k}$ définissent une répresentation linéaire de $\mathfrak{k}$ dans $\mathfrak{a}^p(\mathfrak{m}\ \mathfrak{V})$. Compte tenu de la connexité de K, on voit facilement que la condition (1) est équivalente à la condition

(5) $$(\rho(a) + L_a + \theta(a)) = 0 \quad \text{pour tout} \quad a \in \mathfrak{k} .$$

On a ainsi une nouvelle manière de caractériser le sous-espace $\mathfrak{a}^p_{\mathfrak{k}}(\mathfrak{m}, \mathfrak{v})$ de $\mathfrak{a}^p(\mathfrak{m}, \mathfrak{v})$.

Soit $(e_i)_{i=1,2,\dots N}$ une base orthonormale de $\mathfrak{m}$ pour le produit scalaire sur $\mathfrak{g}$ déduit de la forme de Killing (cf. N° 2) et soit (η_i) la base duale. On choisit d'autre part sur V un produit scalaire vérifiant les conditions (2) et (3) du N. 5 . Avec l'identification faite plus haut, on a

(6) $$d_s = \sum_i \varepsilon(\eta_i) L_{e_i} \qquad d_a = \sum_i \varepsilon(\eta_i)\rho(e_i)$$

(7) $$\partial_s = -\sum_i L(e_i) L_{e_i} \qquad \partial_a = \sum_i L(e_i)\rho(e_i)$$

(8) $$\Delta_s = -\sum_i L^2_{e_i} - \sum_{i,j} \varepsilon(\eta_i) L(e_j) L_{[e_i, e_j]}$$

(9) $$\Delta_a = \sum_i \rho(e_j)^2 + \sum_{i,j} \varepsilon(\eta_i) L(e_j)\rho([e_i, e_j])$$

Ces formules sont valables sur le sous-espace $\mathfrak{a}^n_{\mathfrak{k}}(\mathfrak{m}\ \mathfrak{v})$ de $\mathfrak{a}^p(\mathfrak{m}, \mathfrak{v})$ pour tout p . Elles seront prises pour définition des opérateurs $d_s, d_a, \partial_s, \partial_a, \Delta_s, \Delta_a$, sur le complémentaire.

Soit $(a_k)_{k=1,1,\dots n-N}$ une base orthonormale de $\mathfrak{k}$ pour le produit scalaire sur $\mathfrak{g}$. En remplaçant $[e_i, e_j]$ par $\sum_k \langle e_i, [e_j, a_k]\rangle a_k$ dans (8) et (9) on obtient :

(10) $$\Delta_s = -\sum_i L^2_{e_i} - \sum_k L_{a_k}\,\theta(a_k) ,$$

(11) $$\Delta_a = \sum_i \rho(e_i)^2 + \sum_k \rho(a_k)\,\theta(a_k) .$$

Lemme 1. Pour tout p , on a

$$\Delta = \Delta_a + \Delta_s = \sum_i \rho(e_i)^2 - \sum_k \rho(a_k)^2 - \sum_i L^2_{e_i} + \sum_k L^2_{a_k}$$

sur $\mathfrak{a}^p_k(\mathfrak{m}, \mathfrak{v})$.

L'égalité $\Delta = \Delta_a + \Delta_s$ a été vue au n. 5 . Elle se vérifie facilement en utilisant les expressions (6), (7) . La seconde égalité résulte des (10) , (11) et de (5) .

On remarquera que Δ est la différence des opérateurs de Casimir associés aux représentations linéaires $x \longrightarrow \rho(x)$ et $x \longmapsto L_x$ de $\mathfrak{g}$ dans $\mathcal{A}^p(\mathfrak{m}, \mathcal{V})$.

D'après ce qui a été vu au n. 5 , toute forme harmonique est annulée par Δ_a. Par conséquent, si Δ_a est injectif sur $\mathcal{A}^p(\mathfrak{m}, \mathcal{V})$, alors $H^p(M, E) = H^p(\Gamma, \rho) = (0)$ On va traduire l'injectivité de Δ_a en termes de formes quadratiques.

Pour tout p, soit $\mathcal{A}^p(\mathfrak{m}, V)$ l'espace des formes alternées de degré p sur $\mathfrak{m}$ à valeurs dans V. Il s'identifie de manière évidente à un sous-espace de $\mathcal{A}^p(\mathfrak{m}, \mathcal{V})$ et $\mathcal{A}^p(\mathfrak{m}, \mathcal{V})$ peut être considéré comme l'espace des fonctions différentiables sur $\Gamma \backslash G$ à valeurs dans $\mathcal{A}^p(\mathfrak{m}, V)$. Les opérateurs $\rho(x)$ et $\theta(a)$ pour $x \in \mathfrak{g}$ et $a \in \mathfrak{k}$ laissent stable $\mathcal{A}^p(\mathfrak{m}, V)$. On définit sur $\mathcal{A}^p(\mathfrak{m}, V)$ un produit scalaire (,) en posant

$$(\omega, \omega') = \sum_{\tau} \langle \omega \cdot (e_{\tau 1}, e_{\tau 2}, \dots e_{\tau p}) , \omega'(e_{\tau 1}, e_{\tau 2}, \dots e_{\tau p}) \rangle$$

où τ parcourt l'ensemble des applications strictement croissantes de $[1, p]$ dans $[1, N]$. On voit facilement que

$$(\rho(b)\omega, \omega') = (\omega, \rho(b)\omega') ,$$

$$(\rho(a)\omega, \omega') = -(\omega, \rho(a)\omega') ,$$

$$(\theta(a)\omega, \omega') = -(\omega, \theta(a)\omega') ,$$

quels que soient $a \in \mathfrak{k}$, $b \in \mathfrak{m}$, $\omega, \omega' \in \mathcal{A}^p(\mathfrak{m}, V)$. Cela étant, on définit une forme bilinéaire symétrique $\mathcal{H}_\rho^{(p)}$ sur $\mathcal{A}^p(\mathfrak{m}, V)$ en posant

$$\mathcal{H}_\rho^{(p)}(\omega, \omega') = \Big(\sum_i \rho(e_i)^2 + \sum_k \theta(a_k)\rho(a_k)\Big)\omega, \omega'\Big)$$

quelles que soient $\omega, \omega' \in \mathcal{A}^p(\mathfrak{m}\ V)$.

Théorème 1 (Matsushima-Murakami). Si la forme $\mathcal{H}^{(p)}$ est définie positive, alors $H^p(M,E) = H^p(\Gamma, \rho) = (0)$.

C'est une conséquence immédiate de l'expression (11) de Δ_a.

Remarque. Lorsque Γ est un sous-groupe discret uniforme de G, non nécessairement sans torsion, on a encore $H^p(\Gamma, \rho) = (0)$ si $\mathcal{H}^{(p)}$ est definie positive. On se ramène en effet au cas où Γ est sans torsion en utilisant le Théorème de Selberg cite au n. 1 et la suite spectrale de Hochschild-Serre.

Des conditions suffisantes pour que $\mathcal{H}_\rho^{(p)}$ soit définie positive ont été données par Raghunathan, notamment dans le cas $p = 1$. Le cas où ρ = ad (représentation adjointe de G) et où $p = 1$ avait été traité par A. Weil (24) ; c'est le cas intéressant pour la déformation du sous-groupe discret $\Gamma \subset G$.

Lemme 2 (A. Weil). Si $\mathfrak{g}$ ne contient pas d'idéal de rang 1 et si $\mathfrak{k}$ ne contient pas d'idéal de $\mathfrak{g}$ différent de (0), la forme $\mathcal{H}^{(1)}_{ad}$ est définie positive.

Puisque $\mathfrak{g}$ est somme directe de $\mathfrak{k}$ et de $\mathfrak{m}$, l'espace $\mathcal{A}^1(\mathfrak{m}, \mathfrak{g})$ est somme directe de Hom $(\mathfrak{m}, \mathfrak{k})$ et de Hom$(\mathfrak{m}, \mathfrak{m})$. Ces deux sous-espaces sont orthogonaux pour la forme $\mathcal{H}^{(1)}_{ad}$ car ils sont stables par Δ_a. On montrera d'abord que $\mathcal{H}^{(1)}_{ad}$ est définie positive sur Hom$(\mathfrak{m}, \mathfrak{k})$ ou, ce qui revient au même, que Δ_a est injective sur Hom$(\mathfrak{m}, \mathfrak{k})$. Si $\omega \in$ Hom $(\mathfrak{m}, \mathfrak{k})$ et si $\Delta_a \omega = 0$, alors $d_a \omega = 0$ et ceci implique $[e_i, \omega(e_j)] = 0$ quels que soient i, j d'après (6). On en déduit que $ad(\omega(e_i))\mathfrak{m} = (0)$ pour tout i. Puisque $\mathfrak{k}$ ne contient pas d'idéal de $\mathfrak{g}$ différent de (0), ceci entraîne $\omega(e_i) = 0$ pour tout i et par conséquent $\omega = 0$. Montrons maintenant que $\mathcal{H}^{(1)}_{ad}$ est définie positive sur Hom $(\mathfrak{m}, \mathfrak{m})$. Pour tout $\omega \in$ Hom $(\mathfrak{m}, \mathfrak{m})$, posons $\omega(e_i) = \sum_j \omega_{ij} e_j$. On a alors

$$\mathcal{H}^{(1)}_{ad}(\omega, \omega) = \frac{1}{2} \sum_{i,j} \omega_{ij}^2 + \sum_{i,j,s,r} R_{isjr}\, \omega_{ri}\, \omega_{js},$$

où $R_{isjr} = \langle [[e_r, e_j], e_s], e_i \rangle$; ces coefficients R_{isjr} sont les composantes du tenseur de courbure de la métrique riemannienne symétrique de G/K au point K par rapport à la base de vecteurs consituée par les images des e_i. On a les relations suivantes :

(12) $$R_{irjs} = - R_{rijs} = R_{jsir} ,$$

(13) $$R_{irjs} + R_{ijsr} + R_{isrj} = 0 \text{ (Jacobi-Bianchi)} ,$$

(14) $$\sum_r R_{irjr} = - \frac{1}{2} \delta_{ij} .$$

La relation (14) résulte de la formule (1) du n. 2 . On vérifie en utilisant les relations (12) que le sous-espace des $\omega \in$ Hom $(\mathfrak{m}, \mathfrak{m})$ qui sont symétriques (c.a.d. tels que $\omega_{ij} = \omega_{ji}$ quels que soient i et j) est orthogonal pour la forme $\mathcal{H}^{(1)}_{ad}$ à l'espace des ω antisymétriques. La relation (13) permet de voir que , sur l'espace des ω antisymétriques, la forme quadratique $\sum_{i,j,r,s} R_{irjs} \omega_{si} \omega_{jr}$ est positive et par suite que $\mathcal{H}^{(1)}_{ad}$ est définie positive.

Supposons maintenant que $\omega \in$ Hom $(\mathfrak{m}, \mathfrak{m})$ soit symétrique. On peut choisir la base orthonormale (e_i) de $\mathfrak{m}$ de sorte que chaque e_i soit un vecteur propre de ω . Posons $\omega(e_i) = \lambda_i e_i$ et $r_{ij} = R_{ijji} = \langle [e_i, e_j] , [e_i, e_j] \rangle$ quels que soient i, j . Si $\Delta_a \omega = 0$, on a $\lambda_i + 2 \sum_i r_{ij} \lambda_j = 0$ quel que soit i .

Compte tenu de (14) ceci donne $\sum (\lambda_i + \lambda_j) r_{ij} = 0$. Par conséquent, les r_{ij} étant $\geqslant 0$, si $|\lambda_i| = \lambda = \operatorname{Sup}_j |\lambda_j|$ et si $\lambda_i + \lambda_s \neq 0$, on a $r_{is} = 0$.

On déduit de cela (cf. 24) que, si $\mathfrak{m}_\lambda = \operatorname{Ker}(\omega - \lambda I)$ et $\mathfrak{m}_{-\lambda} = \operatorname{Ker}(\omega + \lambda I)$, et si $\lambda \neq 0$, le sous-espace $\mathfrak{m}_\lambda + \mathfrak{m}_{-\lambda} + [\mathfrak{m}_\lambda, \mathfrak{m}_{-\lambda}]$ de $\mathfrak{g}$ est un idéal produit direct d'idéaux simples de rang 1. Ceci étant exclu, $\lambda = 0$, donc $\omega = 0$ ce qui montre que $\mathcal{H}^{(1)}_{ad}$ est définie positive sur l'espace des éléments symétriques de Hom $(\mathfrak{m}, \mathfrak{m})$.

<u>Théorème 2.</u> (A. Weil) . <u>Soit G une groupe de Lie semi-simple connexe</u>

de centre fini. Si G ne contient ni sous-groupe invariant compact de dim > 0 , ni sous-groupe invariant simple de rang 1, alors pour tout sous-groupe discret uniforme $\Gamma \subset G$, (1) $H^1(\Gamma, \mathrm{ad}) = (0)$.

C'est une conséquence directe du Lemme 2 et du Théorème 1.

Ainsi qu'il a été dit au n. 1 , la relation $H^1(\Gamma, \mathrm{ad}) = 0$ entraîne la trivialite des déformations de Γ sous-les hypothèses du Théorème (cf. [24]) .

n. 7 Nombres de Betti de M

On conserve les hypothèses et les notations du n. 6 . On suppose que $V = \mathbb{R}$, la représentation ρ de G dans V étant alors nécessairement la représentation triviale. L'espace fibré E est alors un espace fibré vectoriel isomorphe au fibré $M \times \mathbb{R}$ et le faisceau $\underline{E}$ s'identifie au faisceau des fonctions numériques localement constantes sur les ouverts de M . Par conséquent, pour tout p l'espace $H^p(M, \underline{E})$ est isomorphe à $H^p(M, \mathbb{R})$. Pour déterminer ces espaces par l'intermédiaire des formes harmoniques, on choisira dans $V = \mathbb{R}$ le produit scalaire multiplication de $\mathbb{R}$. La représentation linéaire ρ de $\mathfrak{g}$ dans V étant nulle, on a $\Delta_a = 0$. D'après le Lemme 1 du n. 6, on a sur $\mathcal{A}_k^p(\mathfrak{m}, \mathcal{V})$:

(1) $$\Delta = \sum_k L_{a_k}^2 - \sum_i L_{e_i}^2 .$$

Par conséquent, pour tout p, $\mathcal{A}_k^p(\mathfrak{m}, \mathbb{R})$ appartient au noyau de Δ . Soit $\mathfrak{J}^p$ l'espace des formes différentielles (scalaires) sur Ω qui sont de degré p et qui sont invariantes par G. On sait qu'en associant à toute forme $\omega \in \mathfrak{J}^p$ sa restriction à l'espace $\mathfrak{m}K \subset T\Omega$, on définit un isomorphis me de $\mathfrak{J}^p$ sur $\mathcal{A}_k^p(\mathfrak{m}, \mathbb{R})$. On a donc en définitive une application linéaire injective canonique de $\mathfrak{J}^p$ dans l'espace des formes différentielles harmoniques de degré p sur M. Cette application peut s'obtenir plus directement comme suit. Puisque $\Omega = G/K$ est un espace riemannien symétrique toute forme différentielle sur G/K qui est invariante par G est une forme harmonique; étant invariante par G elle est composée d'une forme différentielle ω sur $M = \Gamma \backslash G/K$ et de l'étalement $G/K \longrightarrow M$. Puisque cet étalement est une isométrie riemannienne, ω est aussi harmonique.

Les résultats de Matsushima exposés dans la suite donnent une condition suffisante pour que toute forme harmonique de degré p sur M soit image d'une forme différentielle invariante par G sur Ω .

Lemme 1. Soit

$$A = \operatorname*{Inf}_{a \in \mathfrak{k}} \frac{- \operatorname{Tr}_{\mathfrak{m}} \operatorname{ad}(a)^2}{\langle a , a \rangle} .$$

Il existe une base orthornormale (a_k) de $\mathfrak{k}$ telle que $A \langle a_k, a_\ell \rangle \leqslant - \operatorname{Tr}_{\mathfrak{m}} \operatorname{ad}(a_k) \operatorname{ad}(a_\ell)$ quels que soient k et ℓ .

La forme f définie par $f(a, b) = - \operatorname{Tr}_{\mathfrak{m}} \operatorname{ad}(a) \operatorname{ad}(b)$ est une forme bilinéaire symétrique positive sur $\mathfrak{k}$; elle est invariante par la représentation adjointe de K. Les idéaux simples $\mathfrak{a}_1, \mathfrak{a}_2, \ldots \mathfrak{a}_s$ et le centre $\mathfrak{a}_o$ de $\mathfrak{k}$ sont donc deux à deux orthogonaux pour f de même que pour le produit scalaire $\langle , \rangle$. La restriction de f à $\mathfrak{a}_o$ coincide avec $\langle , \rangle$. Pour tout idéal simple $\mathfrak{a}_i$ de $\mathfrak{k}$, il existe un nombre $c_i \geqslant 0$ tel que $f(a, b) = c_i \langle a, b \rangle$ quels que soient $a, b \in \mathfrak{a}_i$. On a $A = \operatorname{Inf}(1, c_1, c_2, \ldots c_s)$. Par suite toute base orthonormale de $\mathfrak{k}$ adaptée à la décomposition de $\mathfrak{k}$ en somme directe des $\mathfrak{a}_i$ vérifie la condition du Lemme .

On remarquera que, si $\mathfrak{k}$ ne contient pas d'idéal de $\mathfrak{g}$ différent de (0) , la représentation linéaire de $\mathfrak{k}$ dans $\mathfrak{m}$ définie par restriction de la représentation adjointe de $\mathfrak{g}$ est une représentation fidèle; dans ce cas $A > 0$.

Lemme 2. Soit w une forme volume sur $\Gamma \backslash G$ invariante par G . Quels que soient $x \in \mathfrak{g}$ et $\omega, \omega' \in \mathfrak{a}^p(\mathfrak{m}, \mathcal{V})$, on a

$$\int_{\Gamma \backslash G} (L_x \omega, \omega') w + \int_{\Gamma \backslash G} (\omega, L_x \omega') w = 0 .$$

En effet $(L_x \omega, \omega') + (\omega, L_x \omega') = L_x . (\omega, \omega')$ (dérivée de la fonction (ω, ω') par rapport au champ L_x) . Puisque w est invariant par G, $(L_x \omega, \omega') w + (\omega, L_x \omega') w$ est la dérivée de Lie de la forme $(\omega, \omega') w$ par rapport à L_x . Son intégrale sur la variété compacte $\Gamma \backslash G$ est donc nulle d'après la formule de Stokes .

Lemme 3 . Soit ω un zéro de Δ dans $\mathcal{A}^p_k(\mathfrak{m}, \mathcal{V})$. Quels que soient $i, j \in [1, N]$, on pose $B_{ij} = L_{e_i}\, L(e_j)\, \omega$. On a

$$\int_{\Gamma\backslash G} \left\{ \frac{A}{p} \sum_{i,j} (B_{ij}, B_{ij}) + 2 \sum_{ijrs} R_{srji}\,(B_{ir}, B_{js}) \right\} w \leqslant 0 .$$

On déduit du Lemme 2 que

$$\int \sum_{i,j} (B_{ij}, B_{ij})\, w = - \int \sum_{i,j} (L^2_{e_i}\, \varepsilon(\eta_j)\, L(e_j)\, \omega , \omega)\, w$$

$$= - p \int \sum_i (L^2_{e_i}\, \omega, \omega)\, w .$$

Puisque ω est harmonique $\sum_i L^2_{e_i}\, \omega = \sum_k L^2_{a_k}\, \omega$. On obtient donc en appliquant de nouveau le Lemme 2 puis le Lemme 1 (en faisant un choix convenable de la base a_k) :

$$\int \sum_{ij} (B_{ij}, B_{ij}) w = p \int \sum_k (L_{a_k} \omega, L_{a_k} \omega) w = p \int \sum_{k,\ell} \langle a_k, a_\ell \rangle (L_{a_k} \omega, L_{a_\ell} \omega) \leqslant - \frac{p}{A} \int \sum_{k,\ell} \mathrm{Tr}_{\mathfrak{m}}\, ad(a_k) ad(a_\ell) (L_{a_k} \omega, L_{a_\ell} \omega)\, w$$

$$= \frac{p}{A} \int \sum_{k,i,\ell} \langle [a_k, e_i] , [a_l, e_i] \rangle (L_{a_k} \omega, L_{a_\ell} \omega)\, w$$

$$= \frac{p}{A} \int \sum_{i,j} (L_{[e_i, e_j]}\, \omega, L_{[e_i, e_j]}\, \omega)\, w .$$

Puisque $\omega \in \mathcal{A}^p_k(\mathfrak{m}, \mathcal{V})$, on a $L_{[e_i, e_j]}\, \omega = - \theta\, ([e_i, e_j])\, \omega$ quels que soient i, j . En remplaçant dans l'intégrale précédente l'un des termes $L_{[e_i, e_j]}\, \omega$ par $- \theta\, ([e_i, e_j])\, \omega =$ $\sum_{r,s} \langle e_r, [[e_i, e_j] , e_s] \rangle \varepsilon(\eta_r)\, L(e_s)\, \omega$, et en appliquant de nouveau le Lemme 2, on la met sous la forme

$$- \frac{A}{p} \int \sum_{ijrs} \langle [[e_i, e_j] , e_r] , e_s \rangle\, (B_{ir}, B_{js})\, w .$$

Théorème 1. (Matsushima). Soit $H_{\mathfrak{g}}^{(p)}$ la forme quadratique sur $\mathfrak{m} \otimes \mathfrak{m}$ définie par

$$H_{\mathfrak{g}}^{(p)}(\xi) = \frac{A}{p} \sum_{i,j} \xi_{ij}^2 + 2 \sum_{ijrs} R_{srji}\, \xi_{ir}\, \xi_{js}$$

pour tout $\xi = \sum_{i,j} \xi_{ij}\, e_i \otimes e_j$. Si $\mathfrak{k}$ ne contient pas d'idéal de $\mathfrak{g}$ différent de 0 et si $H_{\mathfrak{g}}^{(p)}$ est définie positive, alors toute forme harmonique (scalaire) de degré p sur M est image d'une forme différentielle invariante par G sur Ω.

Le lemme 3 montre en effet que, si ω est une forme harmonique de degré p, et si $H_{\mathfrak{g}}^{(p)}$ est définie positive, alors $L_{e_i} L(e_j)\omega = 0$ quels que soient i et j. Il en résulte que $L_{e_i} \omega = 0$ pour tout $i \in [1, N]$. On a donc $L_b \omega = 0$ pour tout $b \in \mathfrak{m}$, et par suite $L_a \omega = 0$ pour tout $a \in [\mathfrak{m}, \mathfrak{m}]$. Puisque $\mathfrak{k}$ ne contient pas d'idéal de $\mathfrak{g}$ différent de 0, $\mathfrak{k} = [\mathfrak{m}, \mathfrak{m}]$ et par conséquent la forme ω prend ses valeurs dans les applications constantes de $\Gamma \backslash G$ dans $\mathbb{R}$, ceci montre que $\omega \in \mathcal{A}_{\mathfrak{k}}^p(\mathfrak{m}, \mathbb{R})$.

Corollaire. Si $\mathfrak{k}$ ne contient pas d'idéal de $\mathfrak{g}$ différent de (0) et si $H_{\mathfrak{g}}^{(1)}$ est définie positive, alors $H^1(M, R) = (0)$.

En effet, toute forme différentielle de degré 1 sur Ω qui est invariante par G est nulle parce que $\mathfrak{m} = [\mathfrak{k}, \mathfrak{m}]$.

L'espace des formes différentielles de degré p sur Ω qui sont invariantes par G est canoniquement isomorphe à l'espace des formes différentielles de degré p invariantes sur une forme compacte de l'espace symétrique Ω. Par suite lorsque les hypothèses du théorème sont vérifiées, le p-ième nombre de Betti de M est égal au p-ième nombre de Betti d'une forme compacte de Ω.

Les valeurs de p pour lesquelles $H_{\mathfrak{g}}^{(p)}$ est définie positive ont été déterminées par Matsushima ([12], [13]) dans le cas où Ω est un domaine borné symétrique et par Kaneyuki et Nagano ([9],[10]) dans le cas général.

n. 8 Le cas des domaines bornés symétriques.

On suppose comme au n. 2 que G est un groupe de Lie semi-simple connexe de centre fini et que K est un sous-groupe compact maximal de G. On suppose de plus donnée sur $\Omega = G/K$ une structure holomorphe invariante par G. La dimension réelle de G/K est alors paire; elle sera notée 2n (contrairement à ce qui avait été fait dans les n. précédents). Soient J le tenseur de la structure holomorphe de Ω et q l'application canonique de G sur Ω. Il existe un endomorphisme j de $\mathfrak{g}$ et un seul tel que $\mathfrak{k} = (0)$, $j\mathfrak{m} \subset \mathfrak{m}$ et $q^T o j = J o q^T$. On a $j(j^2+I) = 0$ et en exprimant que J est le tenseur d'une structure intégrable, on obtient

$$(1) \qquad [ja, jb] = j[ja, b] + j[a, jb] + [a, b] \mod \mathfrak{k}$$

quels que soient $a, b \in \mathfrak{g}$. Compte tenu des inclusions $[\mathfrak{k}, \mathfrak{m}] \subset \mathfrak{m}$ et $[\mathfrak{m}, \mathfrak{m}] \subset \mathfrak{k}$, cette condition est équivalente à la condition

$$(2) \qquad [a, jb] = j[a, b]$$

quels que soient $a \in \mathfrak{k}$ et $b \in \mathfrak{g}$.

Supposons de plus que la métrique riemannienne symétrique sur Ω soit hermitienne. On a alors

$$(3) \qquad \langle ja, b\rangle + \langle a, jb\rangle = 0$$

quels que soient $a, b \in \mathfrak{g}$.

Si B désigne la forme de Killing de $\mathfrak{g}$, quels que soient $a, b \in \mathfrak{m}$ et $c \in \mathfrak{k}$, on a $B([ja, b], c) = -B(b, [ja, c]) = -B(b, j[a, c]) = B(jb, [a, c]) = -B([a, jb], c)$. Par suite $[ja, b] + [a, jb] = 0$.

On en déduit, compte tenu de (2) que

$$(4) \qquad j[a, b] = [ja, b] + [a, jb]$$

quels que soient $a, b \in \mathfrak{g}$. Autrement dit j est une dérivation de $\mathfrak{g}$

Puisque $\mathfrak{g}$ est semi-simple, il existe un élément h_o dans $\mathfrak{g}$ et un seul tel que $ad(h_o) = j$. Il est clair que h_o appartient au centre de $\mathfrak{k}$. On observera que la métrique riemannienne symétrique est non seulement hermitienne, mais kaehlerienne, car toutes les formes différentielles invariantes par G sur Ω sont fermées puisque Ω est un espace homogène symétrique. D'après E. Cartan et Harish-Chandra, Ω est holomorphiquement isomorphe à un domaine borné de $\mathbb{C}^N$ et tout domaine borné symétrique s'obtient ainsi.

Comme au n. 2, on supposera donné un sous-groupe discret uniforme sans torsion Γ dans G. Il existe une structure holomorphe sur la variété $M = \Gamma\backslash\Omega$ et une seule telle que l'application canonique $\Omega \longrightarrow M$ soit un étalement holomorphe. Le complexifié TM^c du fibré TM se décompose en somme directe de deux espaces fibrés vectoriels complexes TM^+ et TM^-. Les éléments de TM^+ sont les vecteurs de type (1, 0), c'est à dire les vecteurs de la forme $u - \sqrt{-1}\, Ju$ où $u \in TM$ et où J désigne le tenseur de la structure holomorphe de M. Les éléments de TM^- sont les vecteurs de type (0, 1), c'est à dire les conjugués des éléments de TM^+.

Comme au n. 2, on désigne par E un espace fibré vectoriel de base M muni d'une structure d'espace fibré associé à Ω qui est définie par une représentation linéaire ρ de Γ dans Gl(V) et une application différentiable $p_\Gamma : \Omega \times V \longrightarrow E$. On suppose de plus que V est un espace vectoriel complexe et que ρ est une représentation linéaire complexe de Γ. Il existe alors dans chaque fibre de E une structure complexe et une seule telle que les repères associés aux points de Ω soient des isomorphismes d'espaces vectoriels complexes. Il existe même sur E une structure d'espace fibré vectoriel holomorphe et une seule telle que $p_\Gamma : \Omega \times V \longrightarrow E$ soit un étalement holomorphe, mais <u>cette structure n'interviendra pas.</u>

Les formes différentielles de degré r sur M à valeurs dans E s'identifient aux sections de l'espace fibré $\Lambda^r(TM)^* \otimes E$ (produit tensoriel sur $\mathbb{R}$) ou encore aux sections de l'espace fibré $\Lambda^r(TM^c)^* \otimes_{\mathbb{C}} E$ qui lui est canoniquement isomorphe. Pour tout couple d'entiers (p, q), soit $\mathcal{A}^{p,q}(M, E)$ l'espace des formes de type (p, q) sur M à valeurs dans E. C'est le sous-espace de $\mathcal{A}^{p+q}(M, E)$ constitué par les formes ω telles que $\omega(u_1, u_2, \ldots, u_{p+q}) = 0$ lorsque $u_1, u_2, \ldots u_{p+q}$ est une suite de vecteurs complexes ayant même origine qui comporte plus de p termes dans TM^+ ou plus de q termes dans TM^-. Si $z_1, z_2, \ldots z_N$ sont des coordonnées locales holomorphes sur M, une forme ω de type (p, q) s'écrit localement

$$\omega = \sum_{\sigma,\tau} \varphi_{\sigma,\tau}\, dz_{\sigma 1} \wedge dz_{\sigma 2} \wedge \ldots dz_{\sigma p} \wedge d\bar{z}_{\tau 1} \wedge d\bar{z}_{\tau 2} \wedge \ldots d\bar{z}_{\tau q},$$

où σ et τ parcourent respectivement l'ensemble des applications strictement croissantes de $[1, p]$ et $[1, q]$ dans $[1, N]$ et où les $\varphi_{\sigma,\tau}$ sont des sections locales de E. Il est clair que pour tout r, $\mathcal{A}^r(M, E)$ est somme directe des sous-espaces $\mathcal{A}^{p,r-p}(M, E)$. On a vu au n. 2 que la structure d'associé à Ω définit dans E une connexion linéaire D de courbure nulle. Le transport parallèle relatif à D est compatible avec la structure complexe des fibres. Pour tout champ de vecteurs X sur M, la dérivation covariante D_X est donc linéaire sur $\mathbb{C}$. L'opérateur de différentiation extérieure défini par D se décompose en $d = d' + d''$ où d' est de degré (1, 0) (c'est à dire applique $\mathcal{A}^{p,q}(M, E)$ dans $\mathcal{A}^{p+1,q}(M, E)$) et où d'' est de degré (0, 1) (c'est à dire applique $\mathcal{A}^{p,q}(M, E)$ dans $\mathcal{A}^{p,q+1}(M, E)$). Ces opérateurs sont linéaires sur $\mathbb{C}$.

On supposera dans la suite que ρ est la restriction à Γ d'une <u>représentation linéaire complexe</u> de G dans V que l'on notera encore ρ. On a vu au n. 3 que E est alors canoniquement muni d'une structure d'espace fibré vectoriel associé à l'espace fibré principal $\Gamma\backslash G$. De plus E est muni d'une seconde connexion linéaire: la connexion symétrique D^s. Puisque $\rho(s)$ est un automorphisme complexe de V pour tout $s \in G$, les repères associés aux différents points de $\Gamma\backslash G$ sont des isomorphismes de l'espace vectoriel complexe V sur les fibres de E. Il en résulte que le transport parallèle relatif à D^s est compatible avec la structure complexe des fibres. L'opérateur D^s_X de dérivation covariante par rapport à un champ de vecteurs sur M est donc linéaire sur $\mathbb{C}$. L'opérateur de différentiation extérieure relatif à D^s se décompose en somme d'un opérateur d'_s de degré (1, 0) et un opérateur d''_s de degré (0, 1). Ces opérateurs sont linéaires sur $\mathbb{C}$.

<u>Lemme</u> 1. On a $(d'_s)^2 = (d''_s)^2 = 0$.

Cela résultera trivialement des expressions de d'_s et d''_s qui seront données au n. suivant.

La forme de courbure de D^s qui est une forme de degré 2 sur M à valeurs dans l'espace fibré vectoriel End(E) est somme de trois composantes $R_{o,2}$, $R_{1,1}$, $R_{2,o}$ de types respectifs (0, 2) (1, 1) et (2, 0)

<u>Lemme</u> 2. <u>On a</u> $R_{2,o} = R_{o,2} = 0$

On a en effet $R_{2,0} \wedge \omega = (d'_s)^2 \omega$ et $R_{0,2} \wedge \omega = (d''_s)^2 \omega$ pour toute forme différentielle ω sur M à valeurs dans E

<u>Lemme</u> 3. <u>Il existe sur E une structure holomorphe</u> (de tenseur J) et <u>une seule vérifiant les deux conditions suivantes :</u>

(a) <u>La projection</u> E $\longrightarrow$ M <u>est holomorphe</u>,

(b) <u>les sections holomorphes de</u> E $\longrightarrow$ M <u>sur un ouvert</u> $U \subset M$ <u>sont les sections différentiables</u> φ <u>sur</u> U <u>telles que</u> $d''_s\varphi = 0$.

Si J est le tenseur d'une structure holomorphe sur E vérifiant les conditions (a) et (b), alors : 1) pour tout vecteur $u \in TE$ qui est horizontal pour la connexion D^s, Ju est horizontal pour D^s,

2) les repères associés aux points de $\Gamma \backslash G$ sont des applications holomorphes de V dans E . Ceci démontre l'unicité . Pour prouver l'existence de la structure holomorphe sur E vérifiant les conditions (a) et (b) on observe d'abord qu'il existe une structure presque complexe de tenseur J sur E et une seule vérifiant la condition 1) et la condition 3) : les repères associés aux points de $\Gamma \backslash G$ et la projection $E \longrightarrow M$ sont des applications presque complexes. On vérifie ensuite que J satisfait à la condition d'intégrabilité

$$N(\mathfrak{X}, \mathfrak{Y}) = [J\mathfrak{X}, J\mathfrak{Y}] - [J\mathfrak{X}, \mathfrak{Y}] - J[\mathfrak{X}, J\mathfrak{Y}] - [\mathfrak{X}, \mathfrak{Y}] = 0$$

quels que soient les champs de vecteurs $\mathfrak{X}$ et $\mathfrak{Y}$ sur E. Puisque N est un tenseur , il suffit de montrer la chose pour une famille de champs de vecteurs qui engendre le module des champs de vecteurs sur E. On se limite donc, d'une part aux champs de vecteurs verticaux dont la restriction à chaque fibre est un champs de vecteurs constant, d'autre part aux relèvements horizontaux de champs de vecteurs sur M . La démonstration s'achève en utilisant d'une part le fait que les dérivations covariantes D^s_X sont linéaires sur $\mathbb{C}$, d'autre part la relation $R_{0,2} = 0$ du Lemme 2.

Remarques.

1) Le Lemme 3 est un cas particulier d'un résultat général sur les espaces fibrés vectoriels de base holomorphe dont les fibres possèdent une structure complexe et qui sont munis d'une connexion linéaire conservant la structure complexe des fibres.

2) Lorsque le groupe G admet une complexifié, la structure holomorphe sur E qui est définie par le Lemme 3 peut être obtenu par d'autres procédés mettant en jeu le "facteur d'automorphie canonique" (cf. [16], [19]) , ou une forme compacte de Ω .

3) La structure holomorphe sur E définie par le Lemme 3 est différente de celle que l'on obtiendrait en considérant E comme associé à Ω.

n.9 . Décompositions du Laplacien dans le cas des domaines bornés symétriques.

On a vu au n.6 que l'on pouvait identifier $\mathfrak{a}^p(M,E)$ et l'espace $\mathfrak{a}^p_{\mathfrak{k}}(\mathfrak{m},\mathfrak{V})$. On va commencer par retrouver dans $\mathfrak{a}^p_{\mathfrak{k}}(\mathfrak{m},\mathfrak{V})$ la notion de type introduite au Numéro précédent.

On désignera par $\mathfrak{g}^c$ l'algèbre de Lie $\mathfrak{g}+\sqrt{-1}\,\mathfrak{g}$ complexifiée de $\mathfrak{g}$. Dans $\mathfrak{g}^c$, les valeurs propres 0, $\sqrt{-1}$ et $-\sqrt{-1}$ de $j=\mathrm{ad}(h_o)$ vont correspondre à trois sous-espaces propres :

$$\mathfrak{k}^c = \mathfrak{k} + \sqrt{-1}\,\mathfrak{k} = \mathrm{Ker}\; j$$
$$\mathfrak{n}_+ = \mathrm{Ker}\,(j-\sqrt{-1})$$
$$\mathfrak{n}_- = \mathrm{Ker}\,(j+\sqrt{-1}) .$$

L'espace $\mathfrak{g}^c$ est somme directe de $\mathfrak{k}^c$, $\mathfrak{n}_+$ et $\mathfrak{n}_-$. On a $\mathfrak{m}^c = \mathfrak{m} + \sqrt{-1}\mathfrak{m} = \mathfrak{n}_+ + \mathfrak{n}_-$, $\bar{\mathfrak{n}}_- = \mathfrak{n}_+$. De plus $[k^c, \mathfrak{n}_+] \subset \mathfrak{n}_+$, $[k^c, \mathfrak{n}_-] \subset \mathfrak{n}_-$, $[\mathfrak{n}_+, \mathfrak{n}_+] = [\mathfrak{n}_-, \mathfrak{n}_-] = 0$ et $[\mathfrak{n}_+, \mathfrak{n}_-] \subset k^c$.

On désignera par $\langle\ ,\ \rangle$ la forme bilinéaire complexe sur $\mathfrak{g}^c$ qui prolonge le produit scalaire sur g déduit de la forme de Killing (cf. n.2). Les sous-espaces $\mathfrak{k}^c$, $\mathfrak{n}_+ + \mathfrak{n}_-$ sont orthogonaux pour ces formes. On a $\langle \mathfrak{n}_+, \mathfrak{n}_+\rangle = \langle \mathfrak{n}_-, \mathfrak{n}_-\rangle = \{0\}$. Soit $(e_i)_{i=1,2,\dots 2N}$ una base orthonormale de $\mathfrak{m}$ telle que $e_{i+N} = je_i$ pour $0 < i \leqslant N$. Soit η_i la base duale. On posera

$$x_i = \frac{1}{2}\,(e_i - \sqrt{-1}\;\, je_i)$$

pour $0 < i \leqslant N$. Les x_i constituent une base de $\mathfrak{n}_+$. Les $\bar{x}_i$ constituent une base de $\mathfrak{n}_-$. On a $\langle x_i, \bar{x}_j\rangle = \delta_{ij}$.
Soit ω_i la forme linéaire sur $\mathfrak{m}^c$ telle que $\omega_i(y) = \langle \bar{x}_i, y\rangle$ pour tout $y \in \mathfrak{m}^c$. Les formes ω_i et leurs conjuguées $\bar{\omega}_i$ constituent la base duale de la base $x_i, \bar{x}_j$ de $\mathfrak{m}^c$. Pour tout $i \in [1, N]$, on a :

$$e_i = \frac{1}{\sqrt{2}}(x_i + \bar{x}_i) \quad , \quad e_{i+N} = \frac{\sqrt{-1}}{\sqrt{2}}(x_i - \bar{x}_i) \ ,$$

$$\eta_i = \frac{1}{\sqrt{2}}(\omega_i + \bar{\omega}_i) \ , \quad \eta_{i+N} = -\frac{\sqrt{-1}}{\sqrt{2}}(\omega_i - \bar{\omega}_i) \ .$$

Lemme 1.

$$\sum_i \operatorname{ad} \bar{x}_i \operatorname{ad} \bar{x}_i + \operatorname{ad} \bar{x}_i \operatorname{ad} x_i = \frac{1}{2} \quad \text{sur } \mathfrak{m}^c \ ,$$

$$\sum_i [x_i, \bar{x}_i] = \frac{1}{2\sqrt{-1}} h_0 \ .$$

La première relation s'obtient en exprimant les e_i au moyen des x_i dans la relation $\sum_{i=1}^{2N} \operatorname{ad}(e_i)^2 = \frac{1}{2}$ sur $\mathfrak{m}$ (cf. n. 2). On en déduit que $\sum_i \operatorname{ad}(x_i)\operatorname{ad}(\bar{x}_i) = \sum_i \operatorname{ad}([x_i, \bar{x}_i]) = \frac{1}{2}$ sur $\mathfrak{n}_+$ et que $\sum_i \operatorname{ad}([x_i, \bar{x}_i]) = -\frac{1}{2}$ sur $\mathfrak{n}_-$. On voit d'autre part que $\sum_i \operatorname{ad}([x_i, \bar{x}_i])$ commute avec ad(a) pour tout $a \in \mathfrak{k}$. Il en résulte que $\operatorname{ad}(\sum_i [x_i, \bar{x}_i]) = \frac{1}{2\sqrt{-1}} \operatorname{ad}(h_0)$, ce qui démontre la seconde relation puisque $\mathfrak{g}$ est semi-simple donc de centre réduit à (0).

Puisque V est un espace vectoriel complexe, l'espace $\mathcal{V}$ des fonctions différentiables sur $\Gamma\backslash G$ à valeurs dans V est lui même un espace complexe. Il existe donc un isomorphisme canonique de l'espace $\mathcal{A}^r(\mathfrak{m}, \mathcal{V})$ des formes alternées de degré r sur $\mathfrak{m}$ à valeurs dans $\mathcal{V}$ sur l'espace $\mathcal{A}^r(\mathfrak{m}^c, \mathcal{V})$ des formes alternées complexes de degré r sur $\mathfrak{m}^c$ à valeurs dans $\mathcal{V}$. On dira qu'une forme $\omega \in \mathcal{A}^r(\mathfrak{m}^c, \mathcal{V})$ est <u>de type</u> (p, q) si $r = p+q$ et si $\omega(x_1, x_2, \dots x_{p+q}) = 0$ toutes les fois que la suite $x_1, x_2, \dots x_{p+q}$ contient plus de p termes dans $\mathfrak{n}_+$ ou plus de q termes dans $\mathfrak{n}_-$. Les formes de type (p, q) constituent un sous-espace vectoriel complexe de $\mathcal{A}^{p+q}(\mathfrak{m}^c, \mathcal{V})$ qui sera noté $\mathcal{A}^{p,q}(\mathfrak{m}^c, \mathcal{V})$. Pour tout entier r, $\mathcal{A}^r(\mathfrak{m}^c, \mathcal{V})$ est somme directe des $\mathcal{A}^{p,q}(\mathfrak{m}, \mathcal{V})$ où $p+q=r$.

Pour tout $x \in \mathfrak{g}$, on a défini au n. 6, des opérateurs $\rho(x)$, L_x et $\theta(x)$ sur $\mathcal{A}(\mathfrak{m}, \mathcal{V})$. Par transport de structure ces opérateurs peuvent être considérés comme des opérateurs dans $\mathcal{A}(\mathfrak{m}^c, \mathcal{V})$.

On a vu au n. 6 qu'il existe un homomorphisme injectif canonique de $\mathfrak{a}^r(M, E)$ dans $\mathfrak{a}^r(\mathfrak{m}, \mathcal{V})$ pour tout r . En composant cet homomorphisme avec l'isomorphisme $\mathfrak{a}^r(\mathfrak{m}, \mathcal{V}) \to \mathfrak{a}^r(\mathfrak{m}^c, \mathcal{V})$ on obtient un homomorphisme canonique de $\mathfrak{a}^r(M, E)$ dans $\mathfrak{a}^r(\mathfrak{m}^c, \mathcal{V})$. Pour tout couple d'entiers (p, q) , l'image de $\mathfrak{a}^{p,q}(M, E)$ par cet homomorphisme est le sous-espace $\mathfrak{a}_{\mathfrak{k}}^{p,q}(\mathfrak{m}^c, \mathcal{V})$ de $\mathfrak{a}^{p,q}(\mathfrak{m}^c, \mathcal{V})$ constitué par les formes ω telles que

(1) $$(\rho(a) + L_a + \theta(a))\,\omega = 0$$

pour tout $a \in \mathfrak{k}$.

Dans la suite, <u>on identifiera tout élément de</u> $\mathfrak{a}^{p,q}(M, E)$ <u>à son image</u> dans $\mathfrak{a}_{\mathfrak{k}}^{p,q}(\mathfrak{m}^c, \mathcal{V})$.

Quels que soient $a, b \in \mathfrak{g}$, on posera $L_{a+\sqrt{-1}b} = L_a + \sqrt{-1}L_b$ et $\rho(a+\sqrt{-1}b) = \rho(a) + \sqrt{-1}\,\rho(b)$, ce qui revient à prolonger la représentation linéaire ρ de $\mathfrak{g}$ dans V à la complexifiée $\mathfrak{g}^c$. Cela étant, les x_i et ω_i étant définis comme au début de ce n. les formules (6) du n. 6 donnent :

$$d_s = \sum_i \varepsilon(\omega_i) L_{x_i} + \sum_i \varepsilon(\bar{\omega}_i) L_{\bar{x}_i} ,$$

(2) $$d'_s = \sum_i \varepsilon(\bar{\omega}_i) L_{x_i} ,$$

(3) $$d''_s = \sum_i \varepsilon(\bar{\omega}_i) L_{\bar{x}_i} ,$$

$$d_a = \sum_i \varepsilon(\omega_i)\rho(x_i) + \sum_i \varepsilon(\bar{\omega}_i)\rho(\bar{x}_i) ,$$

(4) $$d'_a = \sum_i \varepsilon(\omega_i)\rho(x_i) ,$$

(5) $$d''_a = \sum_i \varepsilon(\bar{\omega}_i)\rho(\bar{x}_i) ,$$

Puisque $\mathfrak{n}_-$ est une sous-algèbre de Lie abélienne de $\mathfrak{g}^c$ on a $[L_{\bar{x}_i}, L_{\bar{x}_j}] = 0$ quels que soient $i, j \in [1, N]$.

Par suite $(d'_s)^2 = 0$. On voit de même que $(d''_s)^2 = 0$ ce qui démontre le Lemme 1 du n. 8 .

Comme au n. 6 , on suppose choisi dans V un produit scalaire $\langle\, ,\rangle$ tel que

(6) $\langle \rho(a)v, v'\rangle + \langle v, \rho(a)v'\rangle = 0$

(7) $\langle \rho(b)v, v'\rangle - \langle v, \rho(b)v'\rangle = 0$

quels que soient $v, v' \in V$, $a \in \mathfrak{k}$ et $b \in \mathfrak{m}$. On supposera de plus, ce qui est toujours réalisable, que

(8) $\langle \sqrt{-1}\, v, v'\rangle + \langle v, \sqrt{-1}\, v'\rangle = 0$

quels que soient $v, v' \in V$. Compte tenu de (6) et (8) , chaque fibre de E se trouve munie d'un produit scalaire hermitique. Utilisant ce produit sclaire, on définit comme au n. 4 le adjoints ∂ , ∂_s et ∂_a . Ce sont des opérateurs de dégré total -1 , somme d'une composante de degré (-1, 0) , notée ∂' , ∂'_s , ∂'_a et d'une composante de degré (0, -1) notée ∂'' , ∂''_s , ∂''_a respectivement. En utilisant la base x_i, $\bar{x}_i$ de $\mathfrak{m}^c$, on déduit des relations (7) du n. 6 les formules

$$\partial'_s = -\sum_i L(x_i)L_{\bar{x}_i} \ , \quad \partial''_s = -\sum_i i(\bar{x}_i)L_{x_i} \ ,$$

$$\partial'_a = \sum_i L(x_i)\rho(\bar{x}_i) \ , \quad \partial''_a = \sum_i L(\bar{x}_i)\rho(x_i) \ .$$

Du fait que $[\mathfrak{n}_+, \mathfrak{n}_+] = [\mathfrak{n}_-, \mathfrak{n}_-] = (0)$ on déduit que

$$d'_s\partial''_s + \partial''_s d'_s = \partial'_s d''_s + d''_s \partial'_s = 0 \ .$$

Par suite, le Laplacien $\Delta_s = d_s\partial_s + \partial_s d_s$ se décompose en

$$\Delta_s = \Delta'_s + \Delta''_s \ ,$$

avec

$$\Delta'_s = d'_s\partial'_s + \partial'_s d'_s \ , \quad \Delta''_s = d''_s\partial''_s + \partial''_s d''_s \ .$$

Les opérateurs Δ'_s et Δ''_s conservent le type des formes . Il en est donc de même de Δ_s .

Le produit scalaire sur $\mathcal{A}^r(\mathfrak{m}^c, \mathcal{V})$ étant défini comme au n. 6, on a

$$\langle d'_s\, \omega, \bar{\omega}\rangle = \langle \omega, \partial'_s\, \bar{\omega}\rangle, \langle d''_s \omega, \bar{\omega}\rangle = \langle \omega, \partial''_s\, \bar{\omega}\rangle ,$$

quelles que soient les formes $\omega, \omega' \in \mathcal{A}^r_k(\mathfrak{m}^c, \mathcal{V})$. Par suite $\langle \Delta'_s \omega, \omega\rangle \geq 0$ pour toute forme $\omega \in \mathcal{A}^r_k(\mathfrak{m}^c, \mathcal{V})$ et $\langle \Delta'_s\, \omega, \omega\rangle = 0$ implique $\Delta'_s\, \omega = 0$. De même $\langle \Delta''_s\, \omega, \omega\rangle \geq 0$ et $\langle \Delta''_s\, \omega, \omega\rangle = 0$ implique $\Delta''_s\, \omega = 0$.

On vérifie de même que

$$\Delta_a = \Delta'_a + \Delta''_a ,$$

où

$$\Delta'_a = d'_a\, \partial'_a + \partial'_a\, d'_a \quad , \quad \Delta''_a = d''_a\, \partial''_a + \partial''_a d''_a .$$

Les opérateurs Δ'_a, Δ''_a et Δ_a conservent le type des formes.

On voit comme plus haut que $\langle \Delta'_a\, \omega, \omega\rangle \geq 0$ et que $\langle \Delta'_a \omega, \omega\rangle = 0$ implique $\Delta'_a \omega = 0$. De même $\langle \Delta''_a\, \omega, \omega\rangle \geq 0$ et $\langle \Delta''_a \omega, \omega\rangle = 0$ implique $\Delta''_a \omega = 0$.

<u>Lemme</u> 2. Si $\omega \in \mathcal{A}^r_k(\mathfrak{m}^c, \mathcal{V})$, <u>les conditions suivantes sont équivalentes</u>:

(i) $\Delta\, \omega = 0$,

(ii) $\Delta'_a \omega = \Delta''_a\, \omega = \Delta'_s\, \omega = \Delta''_s\, \omega = 0$.

Puisque le produit scalaire dans V vérifie la condition (7), $\Delta = \Delta_a + \Delta_s$. Par suite Δ conserve le type des formes. Il en résulte que pour tout r, l'espace des formes harmoniques de degré r, c'est à dire le noyau de Δ dans $\mathcal{A}^r_k(\mathfrak{m}^c, \mathcal{V})$, se décompose en somme directe de ses intersections avec les sous-espaces $\mathcal{A}^{p,q}_k(\mathfrak{m}^c, \mathcal{V})$. L'espace $H^r(M, \underline{E})$ est donc somme directe des sous-espaces $H^{p,q}(M, \underline{E})$ avec $p+q = r$ constitués par les classes de cohomologie représentées par une forme harmonique de type (p, q).

J. L. Koszul

Pour expliciter les Laplaciens, on définira pour tout $a \in \mathfrak{k}$ des endomorphismes $\theta_+(a)$ et $\theta_-(a)$ de $\mathcal{Q}^r(\mathfrak{m}^c, \mathcal{V})$ en posant

$$(\theta_+(a)\omega)(x_1, x_2, \dots x_r) = -\sum_j \omega(x_1, x_2, \dots [a, x_j^+], \dots x_r),$$

$$(\theta_-(a)\omega)(x_1, x_2, \dots x_r) = -\sum_j \omega(x_1, x_2, \dots [a, x_j^-], \dots x_r),$$

pour toute suite $x_1, x_2, \dots x_r \in \mathfrak{m}^c$, en désignant par x_j^+ (resp. x_j^-) la composante de x_j dans $\mathfrak{n}_+$ (resp $\mathfrak{n}_-$). On a $\theta(a) = \theta_+(a) + \theta_-(a)$ pour tout $a \in \mathfrak{k}$. Quels que soient $a \in \mathfrak{k}$, $x, y \in \mathfrak{g}^c$, on a

$$[\theta_+(a), L_x] = [\theta_-(a), L_x] = [\theta_+(a), \rho(y)] = [\theta_-(a), \rho(y)] = 0.$$

On notera d'autre part les relations

(9) $$\theta_+(a) = \sum_{i,j} \langle a, [x_j, \bar{x}_i] \rangle \varepsilon(\omega_j) \iota(x_i)$$

(10) $$\theta_-(a) = \sum_{i,j} \langle a, [\bar{x}_j, x_i] \rangle \varepsilon(\bar{\omega}_j) \iota(\bar{x}_i)$$

quels que soit $a \in \mathfrak{k}$. Cela étant, $(a_k)_{k=1,2\dots n-2N}$ désignant une base orthonormale de $\mathfrak{k}$, on obtient :

(11) $$\Delta'_s = -\sum_i L_{\bar{x}_i} L_{x_i} + \sum_{i,j} L_{[\bar{x}_i, x_j]} \varepsilon(\omega_j) \iota(x_i)$$

$$= -\sum_i L_{\bar{x}_i} L_{x_i} - \sum_k L_{a_k} \theta_+(a_k)$$

(12) $$\Delta''_s = -\sum_i L_{x_i} L_{\bar{x}_i} - \sum_k L_{a_k} \theta_-(a_k)$$

(13) $$\Delta'_a = \sum_i \rho(\bar{x}_i) \rho(x_i) + \sum_k \rho(a_k) \theta_+(a_k)$$

(14) $$\Delta''_a = \sum_i \rho(x_i) \rho(\bar{x}_i) + \sum_k \rho(a_k) \theta_-(a_k) .$$

Ces formules vont faire apparaître de nouvelles relations entre les Laplaciens d'indice s et les Laplaciens d'indice a dont le moins qu'on puisse dire est que rien ne permettait de les présager. Il est vraisemblable que ces relations n'ont rien à voir avec la situation homogène et qu'on pourrait les obtenir dans un contexte analogue à celui du n. 5 .

Lemme 3, (Lemme de la folkdance) . Pour tout p , on a

$$\Delta'_s - \Delta''_s = \Delta'_a - \Delta''_a$$

sur $\mathcal{Q}^p_{\mathfrak{k}}(\mathfrak{m}^c, \mathcal{V})$.

En effet, d'après (11) , (12) , (13) et (14) , on a

$$\Delta'_s - \Delta''_s - \Delta'_a + \Delta''_a =$$

$$-\sum_k (L_{a_k} + \rho(a_k))(\theta_+(a_k) - \theta_-(a_k)) + \sum_i (L_{[x_i, \bar{x}_i]} + \rho([x_i, \bar{x}_i])) .$$

Quels que soient a, b $\in \mathfrak{k}$, posons $\theta(a + \sqrt{-1}b) = \theta(a) + \sqrt{-1}\theta(b)$. Pour tout $x \in \mathfrak{k}^c$, on a alors $L_x + \rho(x) = -\theta(x)$ sur $\mathcal{Q}^p_k(\mathfrak{m}^c, \mathcal{V})$.

Compte tenu du Lemme 1, il en résulte que

$$\Delta'_s - \Delta''_s - \Delta'_a + \Delta''_a = \sum_k (\theta_+(a_k)^2 - \theta_-(a_k)^2) - \theta(\sum_i [x_i, \bar{x}_i])$$

$$= \sum_k (\theta_+(a_k)^2 - \theta_-(a_k)^2) - \frac{1}{2\sqrt{-1}} \theta(h_o) .$$

Or $\theta(h_o)\omega = (q-p)\sqrt{-1}\omega$ pour toute forme $\omega \in \mathcal{Q}^{p,q}(\mathfrak{m}^c, \mathcal{V})$. D'autre part, $\sum_k ad(a_k)^2$ étant égal à $-\frac{1}{2}$ sur $\mathfrak{m}$, on en déduit que

$$\sum_k \theta_+(a_k)^2 \omega = -\frac{p}{2}\omega , \qquad \sum_k \theta_-(a_k)^2 \omega = -\frac{q}{2}\omega$$

pour toute forme $\omega \in \mathcal{Q}^{p,q}(\mathfrak{m}^c, \mathcal{V})$.

Théorème 1. Pour tout entier r, on a

$$\Delta = 2(\Delta'_s + \Delta''_a) = 2(\Delta'_a + \Delta''_s)$$

sur $\mathcal{A}_k^r(\mathfrak{m}^c, \mathcal{V})$.

C'est une conséquence immédiate du Lemme 3 et de la relation $\Delta = \Delta_a + \Delta_s$.

Corollaire 1. Si $\omega \in \mathcal{A}_k^r(\mathfrak{m}^c, \mathcal{V})$, les conditions suivantes sont équivalentes:

(a) $\Delta\omega = 0$,

(b) $\Delta'_s\omega = 0$ et $\Delta''_a\omega = 0$,

(c) $\Delta''_s\omega = 0$ et $\Delta'_a\omega = 0$.

C'est une conséquence immédiate du Lemme 1 et du Théorème.

Corollaire 2. Si $\omega \in \mathcal{A}_k^{0,q}(\mathfrak{m}^c, \mathcal{V})$, les conditions suivantes sont équivalentes :

(a) $\Delta\omega = 0$,

(b) $\Delta''_s\omega = 0$ et $\rho(x)\omega = 0$ pour tout $x \in \mathfrak{n}_+$,

(c) $\Delta''_a\omega = 0$ et $L_x\omega = 0$ pour tout $x \in \mathfrak{h}$.

En effet, si ω est de type $(0, q)$, alors

$$\Delta'_a\omega = \sum_i \rho(\bar{x}_i)\rho(x_i)\omega .$$

Ceci montre que $b) \Rightarrow a)$. Compte tenu des propriétés du produit scalaire dans V, on a

$$(\Delta'_a\omega, \omega) = \sum_i (\rho(x_i)\omega, \rho(x_i)\omega,$$

ce qui montre que $(a) \Rightarrow (b)$. On voit de même que $(a) \Leftrightarrow (c)$.

Corollaire 3. Si ω est une forme différentielle de type $(p, 0)$ sur M à valeurs dans E, les conditions suivantes sont équivalentes :

(a) $\Delta\omega = 0$,

(b) $\Delta'_s\omega = 0$ et $\rho(\bar{x})\omega = 0$ pour tout $\bar{x} \in \mathfrak{n}_-$,

(c) $\Delta'\omega = 0$ et $L_{\bar{x}}\omega = 0$ pour tout $\bar{x} \in \mathfrak{h}$.

n. 10 . Sous-fibrés holomorphes de E.

Soit V^o un sous-espace complexe de V stable par $\rho(a)$ pour tout $a \in \mathfrak{k}$, autrement dit un sous-module de V considére comme $\mathfrak{k}^c$-module. Soit $E^o = p_k(\Gamma\setminus G, V^o)$ le sous-espace fibré de E qui correspond à V^o. Il est stable par transport parallèle pour la connexion linéaire symétrique. Il en résulte que E^o est un sous-espace fibré vectoriel holomorphe de E . Dans l'isomorphisme canonique de $\mathcal{A}^{p,q}(M, E)$ sur $\mathcal{A}_{\mathfrak{k}}^{p,q}(\mathfrak{m}^c, \mathcal{V})$, l'espace $\mathcal{A}^{p,q}(M, E^o)$ des formes differentielles de type (p, q) à valeurs dans E^o est appliqué sur le sous espace $\mathcal{A}_{\mathfrak{k}}^{p,q}(\mathfrak{m}^c, \mathcal{V}^o)$ de $\mathcal{A}_{\mathfrak{k}}^{p,q}(\mathfrak{m}^c, \mathcal{V})$ où $\mathcal{V}^o$ designe l'espace des fonctions différentiables sur $\Gamma\setminus G$ à valeurs dans V^o.

Lemme 1. Les sous-espaces $\mathcal{A}_{\mathfrak{k}}^{p,q}(\mathfrak{m}^c, \mathcal{V}^o)$ sont stables par les opérateurs d'_s, d''_s, ∂'_s, ∂''_s, Δ'_s, Δ''_s .

C'est en évidence sur les formules du n. 9 .

En général $\mathcal{A}_{\mathfrak{k}}^{p,q}(\mathfrak{m}^c, \mathcal{V}^o)$ n'est pas stable par les opérateurs d'indice a tels que Δ'_a et Δ''_a car ceux-ci font intervenir des endomorphismes $\rho(x)$ avec $x \in \mathfrak{m}^c$. On va voir qu'il y aura cependant stabilité par Δ'_a et Δ''_a pour des sous-espaces V^o convenables.

Puisque h_o appartient au centre de $\mathfrak{k}$, $\rho(h_o)$ commute avec $L_a + \theta(a) + \rho(a)$ quel que soit $a \in \mathfrak{k}$. Par conséquent $\rho(h_o)$ laisse stable le sous-espace $\mathcal{A}_{\mathfrak{k}}^{p,q}(\mathfrak{m}^c, \mathcal{V})$ de $\mathcal{A}^{p,q}(\mathfrak{m}^c, \mathcal{V})$ quels que soient p et q .

Lemme 2. Quels que soient p, q , on a

$$[\rho(h_o), \Delta'_a] = [\rho(h_o), \Delta''_a] = 0$$

sur $\mathcal{A}^{p,q}(\mathfrak{m}^c, \mathcal{V})$.

Posons $C = \sum_i (\rho(x_i)\rho(\bar{x}_i) + \rho(\bar{x}_i)\rho(x_i)) - \sum_k \rho(a_k)^2$.

C'est l'opérateur de Casimir de la représentation ρ . Par suite C commute avec $\rho(x)$ pour tout $x \in \mathfrak{g}^c$. Puisque h_o appartient au centre de $\mathfrak{k}$, $\rho(h_o)$ commute avec $\rho([x_i, \bar{x}_i]) = \rho(x_i)\rho(\bar{x}_i) - \rho(\bar{x}_i)\rho(x_i)$

quel que soit i . Il en résulte que $\rho(h_o)$ commute avec $\sum_i \rho(x_i)\rho(\bar{x}_i)$ et $\sum_i \rho(\bar{x}_i)\rho(x_i)$, ce qui démontre le Lemme compte tenu des expressions de Δ'_a et Δ''_a données au n. 9 (formules (13) et (14)).

Soit ζ une valeur propre de $\rho(h_o)$ considéré comme endomorphisme de V et soit V_ζ le sous-espace propre noyau de $\rho(h_o) - \zeta I$ dans V. C'est un sous-module de V considéré comme $\mathfrak{k}^c$-module. Il lui correspond donc un sous-fibré vectoriel holomorphe $p_K(\Gamma\backslash G, V_\zeta)$ dans E que l'on notera E_ζ . On note $\mathcal{V}_\zeta$ l'espace des fonctions différentiables sur $\Gamma\backslash G$ à valeurs dans V_ζ . Quels que soient p et q, $\mathcal{A}^{p,q}(\mathfrak{m}^c, \mathcal{V}_\zeta)$ est le noyau de $\rho(h_o) - \zeta I$ considéré comme endomorphisme de $\mathcal{A}^{p,q}(\mathfrak{m}^c, \mathcal{V})$.

Lemme 3. *Pour toute valeur propre* ζ *de* $\rho(h_o)$ *dans* V, *le sous-espace* $\mathcal{A}^{p,q}(\mathfrak{m}^c, \mathcal{V}_\zeta)$ *est stable par* Δ'_a *et* Δ''_a .

C'est une conséquence immédiate du Lemme 2.

Puisque K est un sous-groupe compact de G, toutes les valeurs propres de $\rho(h_o)$ sont imaginaires pures. D'autre part, puisque $ad(h_o) = \sqrt{-1}$ sur $\mathfrak{n}_+$ et $ad(h_o) = -\sqrt{-1}$ on a $\rho(x)V_\zeta \subset V_{\zeta+\sqrt{-1}}$ pour tout $x \in \mathfrak{n}_+$ et $\rho(x)V_\zeta \subset V_{\zeta-\sqrt{-1}}$ pour tout $x \in \mathfrak{n}_-$.

Lemme 4 . *Soient* μ (resp. λ) *la valeur propre de* $\rho(h_o)$ *telle que* $\frac{\mu}{\sqrt{-1}}$ *soit maximum (resp. telle que* $\frac{\lambda}{\sqrt{-1}}$ *soit minimum). Si* ρ *est une représentation simple de* G, *alors* V_μ *est l'espace des* $v \in V$ *tels que* $\rho(x)v = 0$ *pour tout* $x \in \mathfrak{n}_-$.

La démonstration est une variante de celle qui donne les poids d'une représentation simple d'une algèbre de Lie simple de rang 1.

Compte tenu des Cor. 2 et 3 du Théorème 1 du n. 9 , on obtient le résultat suivant :

Théorème 1. *La représentation* ρ *de* $\mathfrak{g}$ *dans* V *étant supposée simple, soit* μ (resp. λ) *la valeur propre de* $\rho(h_o)$ *telle que*

$\mu/\sqrt{-1}$ soit maximum (resp. telle que $\lambda/\sqrt{-1}$ soit minimum). Pour tout entier q, l'espace des zéros de Δ dans $\mathcal{A}_k^{o,q}(m^c, \mathcal{V})$ coincide avec l'espace des zéros de Δ_s'' dans $\mathcal{A}_k^{o,q}(m^c, \mathcal{V}_\mu)$; l'espace de zéros de Δ dans $\mathcal{A}_k^{q,o}(m^c, \mathcal{V})$ coincide avec l'espace des zéros de Δ_s' dans $\mathcal{A}_k^{q,o}(m^c, \mathcal{V}_\lambda)$.

Pour toute valeur propre ζ de $\rho(h_o)$, soit $\underline{E}_\zeta^o$ le faisceau des sections holomorphes de E_ζ. Ce sont les sections de E_ζ annulées par d_s''. Par conséquent, pour tout entier q, $H^q(M, \underline{E}_\zeta^o)$ est l'espace de cohomologie de degé q du complexe

$$\mathcal{A}^{o,o}(M, E_\zeta) \longrightarrow \mathcal{A}^{o,1}(M, E_\zeta) \ldots \mathcal{A}^{o,q}(M, E_\zeta) \xrightarrow{d_s''} \ldots$$

Compte tenu du Théorème de Hodge-Kodaira (cf. [2]), il existe donc un isomorphisme canonique de $H^q(M, \underline{E}_\zeta^o)$ sur le noyau de Δ_s'' dans $\mathcal{A}^{o,q}(M, E_\zeta)$. Plus généralement, soit $\underline{E}_\zeta^p$ le faisceau des formes différentielles holomorphes de degré p sur M à valeurs dans E_ζ ; il existe un isomorphisme canonique de $H^q(M, \underline{E}_\zeta^p)$ sur le noyau de Δ_s'' dans $\mathcal{A}^{p,q}(M, E_\zeta)$. Le Théorème 1 admet donc les corollaires suivants :

Corollaire 1. Pour tout entier q, il existe un isomorphisme canonique de $H^p(M, \underline{E}_\mu^o)$ sur $H^{o,q}(M; \underline{E})$.

Corollaire 2. Pour tout entier q, il existe une application linéaire injective de $H^p(M, \underline{E}_\mu^o)$ dans l'espace des zéros de Δ_a dans $\mathcal{A}_k^{o,q}(m^c, \mathcal{V})$.

On exploitera ce Corollaire au n. suivant pour donner des conditions entraînant la nullité de $H^q(M, \underline{E}^o)$.

Corollaire 3. Pour tout entier p, il existe une application linéaire canonique injective de l'espace $H^{p,o}(M, \underline{E})$ dans l'espace $H^o(M, \underline{E}_\lambda^p)$ des formes différentielles holomorphes de degré p sur M à valeurs dans E_λ.

Si p = N, cette application est un isomorphisme.

La première assertion résulte directement du Théorème 1, . Pour démontrer la seconde on remarque que pour tout $a \in \mathfrak{k}$, $\theta_+(a)$ est nul, sur les formes de type (N, 0) . On a donc $\Delta''_s = \Delta'_s$ sur $\mathcal{A}^{N,0}(\mathfrak{m}^c, \mathcal{V})$.

N. 11 Noyau de Δ_a

On supposera dans ce N° que ρ est une representation simple de $\mathfrak{g}^c$ dans V. L'injection canonique de V dans $\mathcal{V}$ (qui associe à tout élément v la fonction constante égale à v sur $\Gamma\backslash G$) définit une injection de $\mathcal{A}^{p,q}(\mathfrak{m}^c, V)$ dans $\mathcal{A}^{p,q}(\mathfrak{m}^c, \mathcal{V})$. Si F est l'algèbre des fonctions différentiables sur $\Gamma\backslash G$ à valeurs dans $\mathbb{C}$, on obtient un homomorphisme canonique du F-module $F \otimes_{\mathbb{C}} \mathcal{A}^{p,q}(\mathfrak{m}^c, V)$ dans le F-module $\mathcal{A}^{p,q}(\mathfrak{m}^c, \mathcal{V})$. Cet homomorphisme est visiblement un isomorphisme. On identifiera dans la suite l'espace $\mathcal{A}^{p,q}(\mathfrak{m}^c, V)$ à son image dans $\mathcal{A}^{p,q}(\mathfrak{m}^c, \mathcal{V})$; elle est stable par les opérateurs Δ'_a et Δ''_a. Le noyau de Δ_a dans $\mathcal{A}^{p,q}(\mathfrak{m}^c, \mathcal{V})$ est donc le F-module engendré par le noyau de Δ_a dans $\mathcal{A}^{p,q}(\mathfrak{m}^c, V)$.
On va indiquer dans la suite, des résultats dus principalement à B. Kostant, (cf. [11], [16]), qui donnent des renseignements précis sur ce noyau.

Puisque les endomorphismes $\rho(a)$, $\theta_+(a')$ et $\theta_-(a'')$ commutent quels que soient $a, a', a'' \in \mathfrak{k}^c$, $\rho + \theta_+$ et $\rho + \theta_-$ sont des représentations linéaires de $\mathfrak{k}^c$ dans $\mathcal{A}(\mathfrak{m}^c, V) = \bigoplus_r \mathcal{A}^r(\mathfrak{m}^c, V)$.

Lemme 1. Tout sous-module de la représentation $\rho + \theta_+$ de $\mathfrak{k}^c$ dans $\mathcal{A}(\mathfrak{m}^c, V)$ est stable par Δ'_a.

Soit en effet $C = \sum_i (\rho(x_i)\rho(\bar{x}_i) + \rho(\bar{x}_i)\rho(x_i)) - \sum_k \rho(a_k)^2$ l'opérateur de Casimir de la représentation ρ de $\mathfrak{g}^c$ dans $\mathcal{A}(\mathfrak{m}^c, V)$. Puisque V est un $\mathfrak{g}^c$-module simple, C est une homothétie. Soit

$C' = -\sum_k (\rho(a_k) + \theta_+(a_k))^2$ l'opérateur de Casimir de la représentation $\rho + \theta_+$ de $\mathfrak{k}^c$ dans $\mathcal{A}(\mathfrak{m}^c, V)$. D'après la relation (13) du n. 9, on a

$$2\Delta'_a = 2\sum_k \rho(a_k)\,\theta_+(a_k) + C + \sum_k \rho(a_k)^2 + \rho\Big(\sum_i [x_i, \bar{x}_i]\Big)$$

$$= C - C' - \frac{1}{2\sqrt{-1}}\,(\rho(h_o) + \theta(h_o))$$

ce qui démontre le Lemme.

On a un résultat analogue pour $\rho + \theta_-$:

Lemme 2. Tout sous-module de la représentation $\rho + \theta_-$ de $\mathfrak{k}^c$ dans $\mathcal{A}(\mathfrak{m}^c, V)$ est stable par Δ''_a.

On choisira une sous-algèbre abélienne maximale $\mathfrak{h}$ dans $\mathfrak{k}$. Elle contient $\mathfrak{h}$ et $\mathfrak{h}^c = \mathfrak{h} + \sqrt{-1}\,\mathfrak{h}$ est une sous-algèbre de Cartan de $\mathfrak{g}^c$.

Soit R l'ensemble des racines de $\mathfrak{g}^c$ relatives à $\mathfrak{h}^c$. On notera Q^+ l'ensemble des $\alpha \in R$ telles que $\alpha(h_o) = \sqrt{-1}$ et Q^- l'ensemble des $\alpha \in R$ telles que $\alpha(h_o) = -\sqrt{-1}$. Il est clair que $\mathfrak{n}_+$ est somme directe des sous-espace propres $\mathfrak{g}^c_\alpha$ où $\alpha \in Q^+$ et que $\mathfrak{n}_-$ est somme directe des sous-espaces propres $\mathfrak{g}^c_\beta$ où $\beta \in Q^-$. On a visiblement $Q^+ = -Q^-$. L'ensemble $Q = Q^+ \cup Q^-$ est appelé l'ensemble des racines non compactes. L'ensemble $R_k = R - Q$ est appelé l'ensemble des racines compactes ; c'est l'ensemble des racines de $\mathfrak{k}^c$ relatives à $\mathfrak{h}^c$ (qui est aussi une sous-algèbre de Cartan de $\mathfrak{k}^c$).

On choisira un système de racines simples tel que Q^+ soit contenu dans l'ensemble R^+ des racines positives. Pour tout $\alpha \in Q^+$, on choisira un $y_\alpha \in \mathfrak{g}^c_\alpha$ tel que $\langle y_\alpha, \bar{y}_\alpha \rangle = 1$ et on posera $h_\alpha = [y_\alpha, \bar{y}_\alpha]$. Les y_α, pour $\alpha \in Q^+$ constituent une base de $\mathfrak{n}_+$ et $\langle y_\alpha, \bar{y}_\beta \rangle = \delta_{\alpha,\beta}$ quels que soient $\alpha, \beta \in Q^+$. Si $h \in \mathfrak{h}^c$, alors

$$\langle h, h_\alpha \rangle = \langle h, [y_\alpha, y_\alpha] \rangle = \langle [y_\alpha, h], \bar{y}_\alpha \rangle = -\alpha(h) .$$

On a d'autre part, d'après le Lemme 1 du n. 9, $\sum_\alpha [y_\alpha, \bar{y}_{\bar{\alpha}}]$

$$= \frac{1}{2\sqrt{-1}} h_o .$$ Par suite :

$$\sum_{\alpha \in Q^+} h_\alpha = \frac{1}{2\sqrt{-1}} h_o \tag{1}$$

Les h_α appartiennent à $\sqrt{-1}\,\mathfrak{h}$; on posera $\langle \alpha, \beta \rangle = -\langle h_\alpha, h_\beta \rangle$ quels que soient $\alpha, \beta \in R$.

Soit W le groupe de Weyl de $\mathfrak{g}^c$ relatif à $\mathfrak{h}^c$. Pour tout $w \in W$, on posera $\Phi_w = (wR^-) \cap R^+$. On désignera par W^o l'ensemble des $w \in W$ tels que $\Phi_w \subset Q^+$. On sait qu'il existe un élément r dans le groupe de Weyl de $\mathfrak{k}^c$ (donc à fortiori dans le groupe de Weyl W) qui transforme $R_k^+ = R_k \cap R^+$ en $R_k^- = R_k \cap R^-$. Puisque Q^+ est l'ensemble des poids de la représentation de $\mathfrak{k}^c$ dans $\mathfrak{n}_+$ qui s'obtient par restriction de la représentation adjointe, on a $rQ^+ = Q^+$.

On notera P l'ensemble des poids de ρ , λ'(resp. λ) le plus petit (resp. le plus grand) élément de P .

Il est clair que les poids de la représentation θ_+ de $\mathfrak{k}^c$ dans $\mathfrak{a}^{p,q}(\mathfrak{m}^c, V)$ sont de la forme $-\langle A\rangle = -\sum_{\alpha\in A}\alpha$ où A est une partie de Q^+ contenant p éléments . On montre que les poids minimaux des composantes isotypiques de la représentation θ_+ dans $\mathfrak{a}^{p,q}(\mathfrak{m}^c, V)$ sont de la forme $-\langle\varphi_w\rangle$ où $w\in W^o$ et où Card $\Phi_w = p$. De même, les poids de la représentation θ_- de $\mathfrak{k}^c$ dans $\mathfrak{a}^{p,q}(\mathfrak{m}^c, V)$ sont de la forme $\langle A\rangle$ où A est une partie de Q^+ contenant q éléments et les poids minimaux des composantes isotypiques sont de la forme Φ_w où $w \in W^o$ et où Card $\Phi_w = q$. Partant de là on démontre les Lemme suivant :

<u>Lemme 3 . Soit</u> $V_{(\xi)}$ <u>la composante isotypique de poids minimal</u> ξ <u>de la représentation</u> ρ <u>de</u> k^c <u>dans</u> V . <u>Pour que</u> Δ'_a <u>ne soit pas injectif dans</u> $\mathfrak{a}^{p,q}(\mathfrak{m}^c, V_{(\xi)})$, <u>il faut qu'il existe un élément</u> $w \in W^o$ <u>tel que</u> $\xi = w\lambda$ <u>et</u> Card $\Phi_w = p$. <u>Pour que</u> Δ''_a <u>ne soit pas injectif dans</u> $\mathfrak{a}^{p,q}(\mathfrak{m}^c, V_{(\xi)})$, <u>il faut qu'il existe un élément</u> $w' \in W^o$ <u>tel que</u> $r\xi = w'\lambda'$ <u>et</u> Card $\Phi_{w'} = q$.

<u>Théorème 1 . Pour que</u> Δ_a <u>ne soit pas injectif dans</u> $\mathfrak{a}^{p,q}(\mathfrak{m}^c, V)$ <u>il faut qu'il existe des éléments</u> $w, w' \in W$ <u>tels que</u>

(a) $rw\lambda = w'\lambda'$,

(b) Card $\Phi_w = p$, Card $\Phi_{w'} = q$.

<u>Corollaire</u> $-^1$. <u>Soit</u> q_ρ <u>le nombre des racines</u> $\alpha \in Q^+$ <u>telles que</u> $\langle\alpha, \lambda\rangle > 0$. <u>Si</u> $0 \leqslant q < q_\rho$, <u>alors</u> Δ_a <u>est injectif sur</u> $\mathfrak{a}^{0,q}(\mathfrak{m}^c, V)$.

En effet, le seul élément $w \in W^o$ tel que $\Phi_w = \emptyset$ est l'élément neutre de W . Par conséquent, si Δ_a n'est pas injectif sur $\mathfrak{a}^{o,q}(\mathfrak{m}^c, V)$, il existe un $w' \in W^o$ tel que $r\lambda = w'\lambda'$ et Card $\Phi_{w'} = q$. Si $\alpha \in Q^+$ et si $\langle \alpha, \lambda \rangle > 0$ alors $\langle \lambda', (w')^{-1} r\alpha \rangle > 0$. Puisque λ' est le poids minimal de ρ , cette inégalité entraîne $(w')^{-1} r\alpha \in R^-$. Puisque $rQ^+ = Q^+$, on a donc $r\alpha \in \Phi_{w'}$. Par conséquent , $q = \text{Card}\, \Phi_{w'} \geq q_\rho$.

<u>Corollaire 2</u> . <u>Soit</u> p_ρ <u>le nombre des racines</u> $\alpha \in Q^+$ <u>telles que</u> $\langle \lambda', \alpha \rangle < 0$. <u>Si</u> $0 \leq p < p_\rho$, <u>alors</u> Δ_a <u>est injectif sur</u> $\mathfrak{a}^{p,o}(\mathfrak{m}^c, V)$.

Démonstration analogue à celle du Cor. 1 .

<u>Corollaire 3</u> . <u>Si</u> $\langle \lambda, \alpha \rangle > 0$ <u>pour toute racine</u> $\alpha \in R^+$ <u>et si</u> $p+q \neq N$, <u>alors</u> Δ_a <u>est injectif sur</u> $\mathfrak{a}^{p,q}(\mathfrak{m}^c, V)$.

Supposons en effet que Δ_a ne soit pas injectif sur $\mathfrak{a}^{p,q}(\mathfrak{m}^c, V)$. Soient $w, w' \in W^o$ tels que $rw\lambda = w'\lambda'$, $p = \text{Card}\, \Phi_w$ et $q = \text{Card}\, \Phi_{w'}$. On va montrer que Q^+ est réunion disjointe de $r\Phi_w$ et $\Phi_{w'}$. Si $\alpha \in r\Phi_w \cap \Phi_{w'}$, alors $r^{-1}\alpha \in wR^-$, donc $\langle w\lambda, \bar{r}^{-1}\alpha \rangle = \langle \lambda, w^{-1}\bar{r}^{-1}\alpha \rangle < 0$. D'autre part, $\alpha \in w'R^-$, donc $\langle w\lambda, \bar{r}^{-1}\alpha \rangle = \langle w'\lambda', \alpha \rangle = \langle \lambda', (w')^{-1}\alpha \rangle \geq 0$. Ceci est impossible car $\langle \lambda, \beta \rangle \neq 0$ pour toute racine β . On voit donc que $r\Phi_w \cap \Phi_{w'} = \emptyset$ et un raisonnement analogue montre que $Q^+ = r\Phi_w \cup \Phi_{w'}$.

On va maintenant tirer quelques conséquences de ces corollaires mettant en jeu les <u>racines simples non compactes</u>. Lorsque $\mathfrak{g}^c$ est simple, on voit facilement qu'il existe exactement une racine simple non compacte : considéré comme $\mathfrak{k}^c$ -module, n_+ est en effet simple et le poids minimal du $\mathfrak{k}^c$- module n_+ est une racine simple non compacte γ . On voit de plus que tout élément de Q^+ est de la forme $\gamma + \beta$ où β est combinaison linéaire à coefficients entiers ≥ 0 des racines simples compactes. Plus généralement, si s est le nombre des idéaux simples de $\mathfrak{g}^c$, il existe exactement s racines simples non compactes et tout élément de Q^+ est de la forme $\gamma + \beta$ où γ est une racine simple non compacte et où β est combinaison linéaire à coefficients entiers $\geqslant 0$ des racines simples compactes. Puisque λ est le plus grand poids de ρ, $\langle \lambda, \mu \rangle \geqslant 0$ pour toute racine $\mu \in R^+$. Pour que $\langle \lambda, \alpha \rangle > 0$ pour tout $\alpha \in Q^+$ il suffit donc que $\langle \lambda, \gamma \rangle > 0$ pour toute racine simple non compacte. Compte tenu du Cor. 1 on a donc le résultat suivant :

<u>Corollaire 4</u>. <u>Si</u> $\langle \lambda, \gamma \rangle > 0$ <u>pour toute racine simple non compacte</u> γ, <u>alors</u> $q_\rho = N$ <u>et</u> Δ_a <u>est injectif sur</u> $\mathfrak{A}^{0,q}(\mathfrak{m}^c, V)$ <u>lorsque</u> $0 \leqslant q < N$.

Dans le cas où ρ est la représentation adjointe de $\mathfrak{g}^c$, on peut donner explicitement la valeur de q_ρ.

<u>Lemme 4</u>. <u>On suppose que</u> $\mathfrak{g}^c$ <u>est simple et que</u> ρ <u>est la représentation adjointe de</u> $\mathfrak{g}^c$. <u>Si</u> γ <u>est la racine simple non compacte, alors</u>:

(a) $$q_\rho = \frac{1}{\langle \gamma, \gamma \rangle} - 1$$

(b) q_ρ <u>est le nombre des racines</u> $\mu \in R^+$ <u>telles que</u> $\mu + \gamma$ <u>soit une racine</u>.

On peut choisir l'ordre des racines simples de telle sorte que le plus grand poids λ de ad soit dans Q^+. Le poids λ est alors également le plus grand poids de la représentation de $\mathfrak{k}^c$ dans $\mathfrak{n}_+$ obtenue par restriction de ad. Puisque $\mathfrak{n}_+$ est un $\mathfrak{k}^c$-module simple, $r\lambda$ est le poids le plus petit du $\mathfrak{k}^c$-module $\mathfrak{n}_+$ et c'est, par conséquent la racine simple non compacte γ. Pour qu'une racine $\alpha \in Q^+$ vérifie la condition $\langle \lambda, \alpha \rangle > 0$, il faut et il suffit que $\langle \gamma, r\alpha \rangle > 0$. Puisque $rQ^+ = Q^+$, ceci montre que q_ρ est égal au nombre des $\alpha \in Q^+$ telles que $\langle \gamma, \alpha \rangle > 0$, ce qui prouve (b). Toute racine $\alpha \in Q^+$ est de la forme $\alpha + \beta$ où β est combinaison linéaire à coefficients entiers $\geqslant 0$ des racines simples compactes. Par suite, si $\alpha \neq \gamma$, alors $\alpha + \gamma$ et $\alpha - 2\gamma$ ne sont pas des racines. Si $\alpha = \gamma$, on a donc $2\frac{\langle\alpha,\gamma\rangle}{\langle\gamma,\gamma\rangle} = 0$ ou 1. Compte tenu de la relation (1) , il en résulte que

$$q_\rho = \sum_{\alpha \in Q^+} 2\,\frac{\langle\alpha,\gamma\rangle}{\langle\gamma,\gamma\rangle} - 1 = \frac{1}{\langle\gamma,\gamma\rangle} - 1 .$$

Lemme 5. *Si* $\mathfrak{g}^c$ *est simple de rang* > 1 *et si* ρ *est la représentation adjointe de* $\mathfrak{g}^c$, *alors* $q_\rho > 1$.

En effet, il existe une racine simple compacte γ' telle que $\langle\gamma,\gamma'\rangle < 0$ et on a $\gamma + \gamma'$ Q^+ ce qui prouve que $q_\rho > 1$ d'après le Lemme 4 .

n. 12 Théorèmes de Nullité .

Les résultats du Numéro précédent, combinés avec ceux du n. 10 donnent deux types de conditions suffisantes de nullité pour la cohomologie . Les unes concernent la cohomologie de M à coefficients dans le faisceau des sections holomorphes de l'espace fibré vectoriel E_μ . Les autres portent sur la cohomologie de M à coefficients dans le faisceau des sections de E ayant une différentielle covariante nulle, c'est à dire finalement sur la cohomologie de Γ à coefficients dans l'espace de la représentation ρ : ces derniers résultats viennent donc s'ajouter à ceux qui ont été obtenus au n. 6 sans supposer que Ω soit un domaine borné.

Théorème 1 (Calabi-Vesentini) . Si Ω est un domaine borné symétrique irréductible et si θ est le faisceau des sections holomorphes du fibré tangent TM , alors $H^q(M, \theta) = 0$ pour $0 \leq q < \frac{1}{\langle\gamma, \gamma\rangle} - 1$, où γ désigne la racine simple non compacte de l'algèbre de Lie du groupe des automorphismes de Ω .

On applique le Cor. 2 du Th. 1, n. 10 au cas de la représentation adjointe . On a alors $E_\mu = TM$, $\theta = \underline{E}^0_\mu$ et le Théorème résulte donc du Cor. 1 au Th. 1 du n. 11 et du Lemme 4 du n. 11.

Corollaire . Si Ω n'est pas isomorphe au disque unité de $\mathbb{C}$, $H^1(M, \theta) =$ $= (0)$.

Cela résulte du Lemme 5 du n. 11.

On en déduit le Théorème énoncé au n. 1 qui concerne le cas d'un domaine borné symétrique éventuellement réductible.

Théorème 2 (Matsuhima, Murakami) . Soient Ω un domaine borné symétrique, G la composante connexe neutre du groupe des automorphismes holomophes de Ω et Γ un sous-groupe discret uniforme de G. Pour toute *représentation simple de* G, *on a* $H^{p,o}(\Gamma, \rho) = (0)$ (resp. $H^{o,q}(\Gamma, \rho) =$ $= (0)$) *pour* $0 \leqslant p < p_\rho$ (resp. $0 < q \leqslant q$).

Cela résulte directement des Cor. 1 et 2 du Th. 1, n. 11 et des Cor. 3 et 2 du Th. 1 , n. 9 .

BIBLIOGRAPHIE

1. A. Andreotti and E. Vesentini, On deformations of discontinuous groups, Acta Math. 112 (1964), 249-298.

2. W.L. Baily, The decomposition theorems for V-manifolds, Amer. J. Math 78 (1956), 862-888

3. A. Borel, On the curvature tensor of the hermitian symetric manifolds, Ann. of Math. (2) 71 (1960), 508-521.

4. ———, Cchomologie et rigidité d'espaces compacts localement symétriques, Séminaire Bourbaki 16e année, (1963/64), Exp. 265, Secrétariat mathématique, Paris, 1964.

5. E. Calabi, On compact riemannian manifolds with constant curvature, I, Differential geometry, Proc. Sympos. Pure Math. Vol. 3, Amer Math. Soc. Providence, R.I., 1961, pp. 155-180.

6. E. Calabi and E. Vesentini, On compact, locally symmetric Kahler manifolds, Ann. of Math. (2) 71 (1960), 472-507.

7. P. Cartier, Remarks on "Lie algebra cohomology and generalized Borel-Weil theorem" by B. Kostant, Ann. of Math. (2) 74 (1961), 388-390.

8. A. Frohlicher and A. Nijenhuis, A theorem of stability of complex structures. Proc. Nat. Acad. Sci. U.S.A. 43 (1957), 239-241.

9. S. Kaneyuki and T. Nagano, On the first Betti numbers of compact quotient spaces of complex semi-simple Lie groups by discrete subgroups, Sci Papers College Gen. Ed. Univ. Tokyo 12 (1962), 1-11.

10. ———, On certain quadratic forms related to symmetric Riemannian spaces, Osaka Math. J. 14 (1962), 241-252.

11. B. Kostant, Lie algebra cohomology and generalized Borel-Weil theorem, Ann. of Math. (2) 74 (1961), 329-387.

12. Y. Matsushima, On the first Betti number of compact quotient spaces of higher dimensional symmetric spaces, Ann. of Math. (2) 75 (1962), 312-330.

13. ———, On Betti numbers of compact, locally symmetric Riemannian manifolds, Osaka Math. J. 14 (1962), 1- 20.

14. ———, A formula on the Betti numbers of locally symmetric Riemann manifolds (to appear).

15. Y. Matsushima and S. Murakami, On vector bundle valued harmonic forms and automorphic forms on symmetric Riemannian manifolds, Ann. of Math. (2) 78 (1963), 365-416.

16. ———, On certain cohomology groups attached to hermitian symmetric spaces, Osaka J. Math. 2 (1965), 1-35.

17. Y. Matsushima and G. Shimura, On the cohomology groups attached to certain vector valued differential forms on the product of the upper half planes, Ann. of Math. (2) 78 (1963), 417-449.

18. S. Murakami, Cohomologies of vector valued forms on compact locally symmetric Riemann manifolds, Proc. Symp. Vol. 9, algebraic groups and discontinuous subgroups, (1966), pp. 387-393 .

19. ————, Cohomology groups of vector valued forms on symmetric spaces, Lecture Notes, Chicago (1966) .

20. M.S. Raghunathan, On the first cohomology of discrete subgroups of semi-simple Lie groups, Amer. J. Math. 78 (1965), 103-139 .

21. ———— A vanishing theorem for the cohomology of arithmetic subgroups of algebraic groups (to appear).

22. ———— Cohomology of arithmetic subgroups of algebraic groups II (to appear).

23. A. Selberg, On discontinuous groups in higher-dimensional symmetric spaces, Contributions to Function Theory, International Colloquium on Function Theory (Bombay, 1960), pp. 147-164, Tata Institute of Fundamental Research, Bombay, 1960.

24. A. Weil, On discrete subgroups of Lie groups, II, Ann. of Maths. (2) 75 (1962), 578-602 .

25. ————, Remarks on the cohomology of groups, Ann of Math. (2) 80 (1964), 149-157 .

CENTRO INTERNAZIONALE MATEMATICO ESTIVO

(C.I.M.E.)

S. MURAKAMI

PLONGEMENTS HOLOMORPHES DE DOMAINES SYMETRIQUES

Corso tenuto ad Urbino dal 5 al 13 luglio 1967

PLONGEMENTS HOLOMORPHES DE DOMAINES SYMETRIQUES

par

S. MURAKAMI

(Osaka - University)

Le but de ces exposés est de donner une introduction rapide aux travaux [1], [2] de I. SATAKE, en se bornant aux sujets directement liés à la géométrie des domaines symétriques. Pour les propriétés de ces domaines on renvoie au cours de A. KORANYI cité [K].

§ 1. Plongements holomorphes.

Soient D, D' deux domaines hermitiens symétriques ; nous désignerons par G (resp. G') la composante neutre du groupe des automorphismes analytiques de D (resp. D'), composante qui coïncide avec celle du groupe des isométries de D (resp. D'). Nous proposons d'étudier les réalisations de D comme sous-domaine de D' ; pour préciser ce problème, introduisons les notions suivantes:

Définitions: un plongement holomorphe $\rho : D \to D'$ est une injection de D dans D' telle que :

(i) ρ soit une application holomorphe

(ii) ρ soit isométrique

(iii) $\rho(D)$ soit une sous - variété totalement géodésique de D'

Deux plongements holomorphes ρ_1, ρ_2 de D dans D' sont dits équivalents s'il existe $g' \in G'$ tel que $\rho_2 = g' \circ \rho_1$.

Le problème des réalisations de D comme sous-domaine de D' peut alors être interprété comme la classification à équivalence près des plongements holomorphes de D dans D'.

Nous savons que G et G' sont des groupes de Lie semisimples connexes de type non-compact à centre trivial. L'Algèbre de Lie $\mathfrak{g}$ (resp $\mathfrak{g}'$) de G (resp. G') peut être identifiée à l'algèbre de Lie des champs de vecteurs de Killing sur D (resp. D'). Tout plongement holomorphe $\rho : D \longrightarrow D'$ définit un isomorphisme local de G dans G', car pour tout $z \in D$ la symétrie de D' de centre $\rho(z)$ induit de manière canonique la symétrie de centre z dans D. Ainsi la différentielle $d\rho : \mathfrak{g} \longrightarrow \mathfrak{g}'$ d'un plongement holomorphe est injective.

Dès maintenant choisissons un point base $o \in D$ et un point base $o' \in D'$; cela étant fait le sous-groupe d'isotropie K (resp. K') de o (resp. o') est un sous-groupe compact maximal qui permet de procéder aux identifications suivantes :

$$D = G/K \qquad\qquad D' = G'/K'$$
$$o = K \qquad\qquad o' = K'$$

Parallèlement les algèbres de Lie sont victimes de décompositions de Cartan :

$$\mathfrak{g} = \mathfrak{k} + \mathfrak{p} \qquad\qquad \mathfrak{g}' = \mathfrak{k}' + \mathfrak{p}'$$

Les plongements holomorphes $\rho : D \longrightarrow D'$ préservant les points base $\rho(o) = o'$ jouissent des propriétés suivantes :

- comme ρ permute avec les symétries de centre o (resp. o') dans D (resp. D'), on a sur les algèbres de Lie :

(1) $d\rho(\mathfrak{k}) \subset \mathfrak{k}'$ $\qquad$ $d\rho(\mathfrak{p}) \subset \mathfrak{p}'$

- la structure complexe de D = G/K (resp. de D') détermine un élément H_0 - élément appelé Z dans [K] [1)] -

[1)] Nous dirons que H_0 est un H-élément de $\mathfrak{g}$.

(resp. H'_o) dans le centre de $\mathfrak{k}$ (resp. $\mathfrak{k}'$) et ρ étant holomorphe on a

$$(2) \qquad d\rho \circ \mathrm{ad}\ H_o = \mathrm{ad}\ H'_o \circ d\rho \qquad \text{sur } \mathfrak{g}.$$

Les conditions (1) et (2) sont équivalentes à l'unique condition suivant:

$$(H_1) \quad d\rho \circ \mathrm{ad}\ H_o = \mathrm{ad}\ H'_o \circ d\rho \qquad \underline{\text{sur}}\ \mathfrak{g}.$$

Inversement si $d\rho : \mathfrak{g} \to \mathfrak{g}'$ est un homomorphisme injectif qui vérifie (H_1) il lui correspond un unique plongement holomorphe $\rho : D \longrightarrow D'$ tel que $\rho(o) = o'$.

Deux plongements holomorphes ρ_1, ρ_2 de D dans D' préservant les points base o et o' sont dits <u>(k)-équivalents</u> s'il existe $k' \in K'$ tel que $\rho_2 = k' \circ \rho_1$; pour que ρ_1 et ρ_2 soient (k)-équivalents il faut et il suffit que :

$$(3) \qquad d\rho_2 = \mathrm{ad}\ k' \circ d\rho_1 .$$

S. IHARA a montré dans [3] que l'équivalence dans G' revient à la (k)-équivalence pour les plongements holomorphes préservant les points base. Cela ramène la classification des réalisations de D par des sous-domaines de D' à celle des homorphismes injectifs $\mathfrak{g} \to \mathfrak{g}'$ vérifiant (H_1), classés modulo (3).

§ 2. Le problème de la classification.

Dans [1] , SATAKE a traité de manière un peu plus générale le problème posé en fin du § 1; précisons son point de vue à l'aide de la

<u>Définition:</u> une algèbre de Lie semi-simple réelle $\mathfrak{g}$ est dite de <u>type-hermitien</u> si toute composante non-compacte (i.e. idéal simple non-compact) est l'algèbre de Lie associée à un domaine sy-

métrique (une telle algèbre de Lie possède en particulier un H-élément) .

La condition (H_1) et la notion de (k)-équivalence ont un sens pour les homomorphismes d'algèbres de Lie de type hermitien ; le problème traité par SATAKE est celui de la classification au sens du § 1 des homomorphismes entre algèbres de Lie de type hermitien.

Soit $\mathfrak{h}$ (resp $\mathfrak{h}'$) une sous-algèbre de Cartan de $\mathfrak{g}$ (resp. $\mathfrak{g}'$) contenue dans $\mathfrak{k}$ (resp. $\mathfrak{k}'$) ; comme deux quelconques de ces sous-algèbres de Cartan sont conjuguées par un ad k pour $k \in K$, tout homomorphisme $d\rho : \mathfrak{g} \rightarrow \mathfrak{g}'$ est (k)-équivalent à un homomorphisme tel que

(4) $$d\rho(\mathfrak{h}) \subset \mathfrak{h}'$$

Enfin on remplace la condition (H_1) par la condition

(H_2) $$d\rho(H_o) = H'_o$$

qui l'implique, ce qui nous permet d'énoncer le résultat suivant :

<u>Proposition 1</u> ([1], [3]) Soient $\mathfrak{g}$, $\mathfrak{g}'$ des algèbres de Lie de type hermitien ; pour tout homomorphisme $d\rho : \mathfrak{g} \rightarrow \mathfrak{g}'$ vérifiant (H_1) il existe une sous-algèbre semi-simple $\mathfrak{g}'' \subset \mathfrak{g}'$ de type hermitien telle que :

(i) $d\rho(g) \subset g''$

(ii) $d\rho : \mathfrak{g} \rightarrow \mathfrak{g}''$ vérifie (H_2)

Indiquons les grandes lignes de la démonstration. On peut supposer que $d\rho$ satisfait à (4) et alors $d\rho(H_o) - H'_o \in \mathfrak{h}'$ et sont centralisateur $\mathfrak{z}$ est une algèbre de Lie réductive. La condition (H_1) signifie que $d\rho(\mathfrak{g}) \subset \mathfrak{z}$. Puisque $\mathfrak{g}$ est semi-simple, $d\rho(\mathfrak{g})$ est contenu dans la partie semi-simple $\mathfrak{g}''$ de $\mathfrak{z}$. On termine en montrant que $\mathfrak{g}''$ satisfait aux conditions (i) et (ii) .

A l'aide de cette proposition SATAKE a obtenu [1] une classification complète dans le cas où $\mathfrak{g}'$ est une algèbre de Lie simple associée à un domaine symétrique irréductible classique. Enfin il a montré que l'on peut alors ramener le problème au cas où D est aussi un domaine irréductible et D' de type $(I)_{p',q'}$ $d\rho$ étant une représentation absolument irréductible vérifiant (H_2) de l'algèbre de Lie $\mathfrak{g}$ dans l'algèbre de Lie du groupe des automorphismes de la forme quadratique de signature (p',q') . En ce cas la condition (H_2) s'exprime sur les poids dominants de $d\rho$, ce qui détermine les représentations $d\rho$ (voir tableau en annexe) .

Dans le cas où $\mathfrak{g}'$ est associée à un domaine irréductible exceptionnel, IHARA [3] a résolu le problème en utilisant la machinerie des racines.

§3. Correspondace entre les composantes des frontières.

Etant donnés deux domaines symétriques D et D', munis de points base respectifs $o \in D$, $o' \in D'$, nous nous proposons d'étudier la correspondance entre les composantes des frontières des compactifiés naturels $\overline{D}$ et $\overline{D'}$.

Supposons connu un groupe de Lie semi-simple connexe G (resp. G') opérant transitivement dans D (resp. D') par automorphismes holomorphes et tel que le sous-groupe d'isotropie K (resp. K') de o (resp. o') dans G (resp. G') soit un sous-groupe compact maximal. Il est possible de modifier la notion de plongement holomorphe de D dans D' à cette situation :

Définition : un plongement holomorphe ρ: $D \longrightarrow D'$ est une application holomorphe de D dans D' qui provient d'un homomorphisme global, noté encore ρ , de G dans G' par passages aux quotients.

Cela revient à dire que l'homomorphisme $\rho : G \to G'$ est tel que $\rho(K) \subset K'$. ρ étant holomorphe, la différentielle $d\rho : \mathfrak{g} \to \mathfrak{g}'$ satisfait à la condition (H_1).

Les structures complexes de D et D' définissent des décompositions des algèbres de Lie $\mathfrak{g}^{\mathbb{C}}$ et $\mathfrak{g}'^{\mathbb{C}}$ complexifiées de et $\mathfrak{g}'$ respectivement :

$$\mathfrak{g}^{\mathbb{C}} = \mathfrak{k}^{\mathbb{C}} + \mathfrak{p}^{+} + \mathfrak{p}^{-} \qquad \mathfrak{g}'^{\mathbb{C}} = \mathfrak{k}'^{\mathbb{C}} + \mathfrak{p}'^{+} + \mathfrak{p}'^{-}$$

(les notations adoptées étant celles de [K]) et la condition (H_1) implique

$$d\rho\,(\mathfrak{p}^{\pm}) \subset \mathfrak{p}'^{\pm}, \quad d\rho\,(\mathfrak{k}^{\mathbb{C}}) \subset \mathfrak{k}'^{\mathbb{C}}.$$

Le plongement d'HARISH-CHANDRA décrit dans [K] chap. 2 permet de réaliser D (resp. D') comme un domaine borné de $\mathfrak{p}^{+}$ (resp. $\mathfrak{p}'^{+}$) ; par définition même de ce plongement on a le résultat suivant :

<u>Proposition 2:</u> avec les notations ci-dessus, le diagramme

$$\begin{array}{ccc} D & \xrightarrow{\ \rho\ } & D' \\ \downarrow & & \downarrow \\ \mathfrak{p} & \xrightarrow{\ d\rho\ } & \mathfrak{p}' \end{array}$$

$d\rho$ complexifiée de la différentielle de ρ

est commutatif.

Autrement dit si l'on identifie D (resp D') à une partie de $\mathfrak{p}^{+}$ (resp. $\mathfrak{p}'^{+}$), on peut considérer ρ comme la restriction à D de l'application $\mathbb{C}$-linéaire $d\rho$, ce qui permet de la prolonger de manière évidente aux compactifications naturelles $\overline{D} \subset \mathfrak{p}^{+}$ et $D' \subset \mathfrak{p}'^{+}$.

Comme toute composante de la frontière de $\overline{D}$ est une composante par arcs holomorphes de $\overline{D}$ ([K], chap. 3), on a

Corollaire : (i) pour toute composante F de la frontière de D il existe une unique composante F' de la frontière de D' telle que $\rho(F) \subset F'$

(ii) si ρ est injective, l'application $F \longmapsto F'$ est injective.

Si o_F est l'unique point de F limite d'une géodésique γ issue de o, on a nécessairement $\rho(o_F) = o'_{F'}$, où $o'_{F'}$ est limite de la géodésique image de γ ([4], théorème 4.14). Cela détermine complétement F' qui est composante connexe par arcs holomorphes de $o'_{F'}$.

Introduisant la conjuguaison $\begin{matrix} \mathfrak{g}^{\mathbb{C}} \longrightarrow \mathfrak{g}^{\mathbb{C}} \\ X \longmapsto \overline{X} \end{matrix}$ relative à $\mathfrak{g}$ on peut expliciter une géodésique ayant o_F pour point limite :

$$\gamma : \lambda \longmapsto \exp(\lambda X_F).o \quad \text{pour } X_F = o_F + \overline{o}_F .$$

§4. L'application de Hermann.

Choisissons maintenant une sous-algèbre de Cartan $\mathfrak{a}$ de la paire symétrique $(\mathfrak{g}, \mathfrak{k})$ et un système fondamental Δ de racines relatives à $\mathfrak{a}$ (voir [K]). Les notions que nous allons introduire ne sont pas liées intrinsèquement au domaine D, mais dépendent du triple $(o, \mathfrak{a}, \Delta)$.

Tout sous-ensemble $\Lambda \subset \Delta$ définit une composante F_Λ de la frontière de $\overline{D}$ selon un procédé canonique décrit dans [K] chap. 3.

Examinons préalablement un exemple : le disque unité $U = \{ z \in \mathbb{C} \mid |z| < 1 \}$ dans lequel on fait opérer par homographies le groupe.

$$G_o = \left\{ \begin{pmatrix} a & b \\ c & d \end{pmatrix} = M \in \underset{=}{M}_2(\mathbb{C}) \;\middle|\; M^* \begin{pmatrix} 1 & o \\ o & -1 \end{pmatrix} M = \begin{pmatrix} 1 & o \\ o & -1 \end{pmatrix},\ \det M = 1 \right\}$$

l'action étant

$$G_o \times U \longrightarrow U$$

$$\left(\begin{pmatrix} a & b \\ c & d \end{pmatrix}, \ z \right) \longmapsto (az + b)(cz + d)^{-1}.$$

L'algèbre de Lie $\mathfrak{g}_o$ de G_o est formée des (2x2)- matrices complexes $\begin{pmatrix} x & y \\ \bar{y} & \bar{x} \end{pmatrix}$ où $\bar{x} = -x$; sa complexifiée $\mathfrak{g}_o^{\mathbb{C}}$ admet pour base

$$H = \begin{pmatrix} 1 & o \\ o & 1 \end{pmatrix}, \quad E^+ = \begin{pmatrix} o & 1 \\ o & o \end{pmatrix}, \quad E^- = \begin{pmatrix} o & o \\ 1 & o \end{pmatrix}$$

et $\mathfrak{g}_o$ est le sous-espace réel engendré par

$$\sqrt{-1} \cdot H, \quad X_o = E^+ + E^-, \quad Y_o = \sqrt{-1} \cdot (E^+ - E^-).$$

La sous-algèbre $\mathfrak{k}_o \subset \mathfrak{g}_o$, correspondant à l'origine $0 \in \mathbb{C}$ choise comme point base de U, est engendrée par $\sqrt{-1} \cdot H$ et l'on peut prendre pour sous-algèbre de Cartan $\mathfrak{a}_o$ de la paire $(\mathfrak{g}_o, \mathfrak{k}_o)$ la sous algèbre engendrée par X_o. Le H-élément dans $\mathfrak{k}_o$ est $1/2 \cdot \sqrt{-1}\,.H$.

On vérifie que l'ensemble réduit à $1 \in \overline{U}$ est une composante de la frontière de $\overline{U}$ et qu'il correspond à $E^+ \in \mathfrak{p}^+$ dans la réalisation d'HARISH-CHANDRA de U ; on a ainsi

$$o_{\{1\}} = E^+ \qquad\qquad X_{\{1\}} = X_o$$

et une géodésique donnant $o_{\{1\}}$ est

$$\gamma : \lambda \longmapsto \exp\ (\lambda X_o) \,.\, 0 = \begin{pmatrix} ch\lambda & sh\lambda \\ sh\lambda & ch\lambda \end{pmatrix} \cdot 0 = th\lambda.$$

Puisque $\lim_{\lambda = \infty} th\lambda = 1$. Cet exemple de domaine symetrique nous fournit le résultat suivant :

<u>Proposition 3:</u> Si $r = \dim \mathfrak{a}$ est le rang de D , il existe un plongement holomorphe :

$$U^r = U \times \ldots \times U \xrightarrow{\ \varkappa\ } D$$

tel que $\varkappa(0, 0, \ldots, 0) = o$ et $d\varkappa(\mathfrak{m}_o^r) \subset \mathfrak{m}$.

De plus ce plongement est unique à un automorphisme de la forme :

$$\begin{array}{lll} U^r \longrightarrow U^r & & \varepsilon_i = \pm 1 \\ (z_i) \longmapsto (\varepsilon_i z_{\sigma(i)}) & & \sigma \in \mathfrak{S}_r \end{array}$$

pres.

On dit que $\varkappa$ est l'application de Hermann (relative à $(o, \mathfrak{m})$) de U^r dans D.

Pour démontrer l'existence du plongement il faut exhiber un homomorphisme global $\rho : G_o^r \longrightarrow G$ passant aux quotients; il se construit à l'aide des vecteurs de $\mathfrak{m}$ correspondant à un système maximal de racines fortement orthogonales. L' "unicité" résulte de la proposition suivante :

<u>Proposition 4:</u> Tout plongement holomorphe $\mu : U^s \longrightarrow D$ tel que $\mu(0, \ldots, 0) = o$ et que $d\mu(\mathfrak{m}_o^s) \subset \mathfrak{m}$ se factorise à l'aide d'un plongement holomorphe $\rho_o : U^s \longrightarrow U^r$ via l'application de Hermann :

$$\begin{array}{ccc} U^s & \xrightarrow{\ \rho_o\ } & U^r \\ & \searrow^{\mu} & \downarrow \varkappa \\ & & D \end{array}$$

<u>Corollaire :</u> Soit $\rho : D \longrightarrow D'$ un plongement holomorphe tel que $\rho(o) = o'$ et $d\rho(\mathfrak{m}) \subset \mathfrak{m}'$; il existe un unique plongement holomorphe $\rho_o : U^r \longrightarrow U^{r'}$ tel que le diagramme

$$\begin{array}{ccc} U^r & \xrightarrow{\ \rho_o\ } & U^{r'} \\ \downarrow \varkappa & & \downarrow \varkappa' \\ D & \xrightarrow{\ \rho\ } & D' \end{array}$$

soit commutatif.

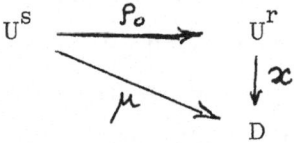

C'est une conséquence de la propriété universelle de l'application de Hermann.

L'application de Hermann fournit des renseignements de nature géométrique sur D (resp D') et sa frontière.

Pour que D soit un domaine tubulaire il faut et il suffit que son application de Hermann vérifie la condit on (H_2). En effet l'image par $\varkappa$ du H-élément $(\frac{\sqrt{-1}}{2} H, \ldots, \frac{\sqrt{-1}}{2} H)$ de $\mathfrak{g}_0^r$ est le H-élément noté Z_1 dans [K]- d'une sous-algèbre semi-simple de $\mathfrak{g}$ ne coincidant avec celle - ci que pour les domaines tubulaires.

Pour tout sous-ensemble $I \subset \{1, \ldots, r\} = R$ désignons par F_I la composante de la frontière de U^r contenant dans le $i^{\text{ème}}$ facteur :

$$\{1\} \text{ si } i \in I \quad , \quad U \text{ tout entier si } i \notin I .$$

Par la proposition 2 on sait qu'il correspond à F_I une unique composante de la frontière de D qui est de la forme F_Λ (voir [K]) pour $\Lambda \subset \Delta$. L'application $F_I \longmapsto F_\Lambda$ est une surjection sur l'ensemble des classes de K-équivalence des composantes de la frontière de D.

Si l'on considère maintenant un plongement holomorphe $\rho : D \to D'$, on peut le relever en un plongement holomorphe $\rho_0 : U^r \to U^{r'}$ en vertu du corollaire de la proposition 4. Comme ρ_0 définit une application entre les composantes des frontières de U^r et $U^{r'}$, on déduit de $F_I \longmapsto F'_{I'}$ une application $I \longmapsto I'$ de l'ensemble des parties de R dans l'ensemble des parties de $R' = \{1, \ldots, r'\}$. La restriction de cette dernière aux parties de la forme $\{i\}$ pour $i \in R$ donne une application $\begin{array}{l} R \longrightarrow \mathfrak{P}(R') \\ i \longmapsto I'_i \end{array}$ et l'on peut montrer que la famille $(I'_i)_{i \in R}$ est disjointe et que $I' = \bigsqcup_{i \in R} I'_i$ quel que soit $I \in \mathfrak{P}(R)$.

Quitte à modifier l'ordre des facteurs dans $U^{r'}$ on peut supposer que si

$I = \{1, \dots, i_o\}$ alors $I' = \{1, \dots, j_o\}$; de plus si D et D' sont irréductibles $j_o = m i_o$, l'entier $m \in \mathbb{N}$ ne dépendant que de ρ (voir tableau en annexe).

§ 5. Le grand théorème.

Pour achever ces exposés signalons que la partie sérieuse du travail de SATAKE [2] commence lorsqu'on prend pour G et G' 1) de groupes algébriques définis sur $\mathbb{Q}$, pour ρ: $G \rightarrow G'$ un homomorphisme rationnel défini sur $\mathbb{Q}$ et que l'on se donne des sous-groupes arithmétiques $\Gamma \subset G$, $\Gamma' \subset G'$ tels que $\rho(\Gamma) \subset \Gamma'$. Il s'agit alors en passant par les résultats suivants :

Lemme 1 ([2], cor. de la prop. 5) ρ envoie les composantes rationnelles de la frontière de D sur les composantes rationnelles de la frontière de D' .

Lemme 2 ([5]) : Soit D^* (resp. D'^*) la réunion de D (resp. D') et de ses composantes rationnelles; l'espace quotient $\Gamma \backslash D^*$ (resp. $\Gamma' \backslash D'^*$) admet une structure analytique naturelle isomorphe à celle d'un sous-espace analytique compact normal d'un espace projectif $\mathbb{P}^N$ (resp. $\mathbb{P}^{N'}$)

de démontrer le grand théorème suivant :

THEOREME : On peut choisir $\mathbb{P}^N$ (resp. $\mathbb{P}^{N'}$) et le plongement $i : \Gamma \backslash D^* \rightarrow \mathbb{P}^N$ (resp. $i' : \Gamma' \backslash D'^* \rightarrow \mathbb{P}^{N'}$) tel qu'il existe une application projective $\tilde{\rho} : \mathbb{P}^N \rightarrow \mathbb{P}^{N'}$ rendant commutatif le diagramme :

$$\begin{array}{ccccc} \Gamma \backslash D & \subset & \Gamma \backslash D^* & \xrightarrow{\ i\ } & \mathbb{P}^N \\ \bar{\rho} \downarrow & & \downarrow & & \downarrow \tilde{\rho} \\ \Gamma' \backslash D' & \subset & \Gamma' \backslash D'^* & \xrightarrow{\ i'\ } & \mathbb{P}^{N'} \end{array}$$

1) La composant neutre.

<u>Corollaire</u> : L'application $\bar{\rho} : \Gamma \backslash D \rightarrow \Gamma' \backslash D'$, induite par ρ se prolonge en un morphisme $\Gamma \backslash D^{*} \rightarrow \Gamma' \backslash D'^{*}$ de variétés algébriques .

<u>Remarque</u> : Dans [2] le grand théorème ci-dessus est énoncé de manière un peu plus faible, mais la forme donnée ici et sa démonstration se trouveront dans un article de SATAKE actuellement en préparation.

Pour terminer l'orateur tient à remercier M. D. Amiguet d'avoir bien voulu lui servir de scribe et ce dernier se résigne à mentionner ce détail .

ANNEXE : <u>Représentations absolument irréductibles $d\rho : \mathfrak{g} \rightarrow (I)_{p',q'}$ et nombre m associé.</u>

$\mathfrak{g}$		$d\rho$	p'	q'	m
$(I)_{p,q}$	$p \geq q \geq 2$	$1_{\mathfrak{g}}$	p	q	1
		I	q	p	
	$p \geq q \geq 1$	Représentation sur l'espace des tenseur antisymétriques de degré m $(1 \leq m \leq p)$	$\binom{p}{m}$	$\binom{p}{m-1}$	$\binom{p-1}{m-1}$
$(II)_p$	$p \geq 5$	$1_{\mathfrak{g}}$	p		2
$(III)_p$	$p \geq 1$	$1_{\mathfrak{g}}$	p		1
$(IV)_p$	$p \geq 4$ pair	Représentation spinorielle	$2^{\frac{1}{2}p-1}$		$2^{\left[\frac{1}{2}(p-3)\right]}$
	$p \geq 1$ impair	Représentation spinorielle	$2^{\frac{1}{2}(p-1)}$		

S. Murakami

BIBLIOGRAPHIE

[1] I. SATAKE : Holomorphic imbeddings of symmetric domains into a Siegel space Am. J. of Math 87 (1965) 425-461

[2] ----------: A note on holomorphic imbeddings and compactifications of symmetric spaces (à paraıtre dans Am. J. of Math)

[3] S. IHARA : Holomorphic imbeddings of symmetric domains into a symmetric domain Proc. of Japan Acad. 42 (1966) 193-197

[4] J.A. WOLF:- A. KORANYI Generalized Caley transformations of bounded symmetric domains Am. J. of Math 87 (1965) 899-939

[5] W.L. BAILY jr - A. BOREL Compactification of arithmetic quotients of bounded symmetric domains (à paraître)

[K] A. KORANYI : Cours au C.I.M.E. 1967 (III session)

CENTRO INTERNAZIONALE MATEMATICO ESTIVO

(C. I. M. E.)

S. MURAKAMI

"FACTEURS D'AUTOMORPHIE ASSOCIES A UN ESPACE HERMITIEN SYMETRIQUE"

Corso tenuto ad Urbino dal 5 al 13 luglio 1967

FACTEURS D'AUTOMORPHIE ASSOCIES A UN ESPACE HERMITIEN SYMETRIQUE.

par

S. MURAKAMI

Soit Ω une variété complexe munie d'un groupe de Lie H opérant à gauche par automorphismes holomorphes. Pour tout espace vectoriel complexe V on considérera le groupe linéaire GL (V) comme une sous-variété ouverte de $\mathbb{C}^{(\dim V)^2}$, ce qui permettra en particulier de parler de fonctions différentiables ou holomorphes à valeurs dans GL(V) .

<u>Définition</u> : un <u>facteur d'automorphie</u> j pour le H-espace à valeurs dans GL(V) est une application différentiable $j : H \times \Omega \to GL(V)$ telle que :

(i) l'application partielle :

$$\Omega \longrightarrow GL(V)$$
$$x \longmapsto j(s, x)$$

soit holomorphe pour tout $s \in H$

(ii) $j(ss', x) = j(s, s'x)\, j(s', x)$ quels que soient $x \in \Omega$ et s, s' dans H.

<u>Remarques</u>

1) Dans la pratique il arrivera fréquemment que H soit un sous-groupe d'un groupe de Lie G transitif dans Ω et que j soit la restriction à H d'un facteur d'automorphie défini sur G tout entier .

2) Si $h : GL(V) \to GL(V')$ est un homomorphisme de groupes et si j est un facteur d'automorphie du H-espace Ω à valeurs dans GL(V) , $h \circ j$ en est un à valeurs dans GL (V') .

<u>Exemples.</u>

1) Si $\rho : H \to GL(V)$ est une représentation linéaire de H dans V,

$$j : \begin{array}{ccc} H \times \Omega & \longrightarrow & GL(V) \\ (s,x) & \longmapsto & \rho(s) \end{array}$$

est un facteur d'automorphie.

2) Si $\Omega \subset \mathbb{C}^m$ est une sous-variété ouverte, le jacobien (complexe) :

$$j : \begin{array}{ccc} H \times \Omega & \longrightarrow & GL(m, \mathbb{C}) \\ (s,x) & \longmapsto & dL_{s,x} \end{array}$$

où $dL_{s,x}$ est la différentielle de $y \longmapsto sy$ au point x, est un facteur d'automorphie, de même que :

$$\begin{array}{ccc} H \times \Omega & \longrightarrow & \mathbb{C}^* \\ (s,x) & \longmapsto & \det(dL_{s,x}) \end{array}$$

en vertu de la remarque 2) .

3) Si $\Omega = \{ z \in \mathbb{C} \mid \operatorname{Im} z > 0 \}$ est le demi-plan de Poincaré dans lequel

$$G = SL(2, \mathbb{R}) = \left\{ \begin{pmatrix} a & b \\ c & d \end{pmatrix} \in M_2(\mathbb{R}) \,\middle|\, ad - bc = 1 \right\}$$

opère par homographies, un calcul simple montre que $dL_{s,x} = (cx+d)^{-2}$ pour $s = \begin{pmatrix} a & b \\ c & d \end{pmatrix}$ et $x \in \Omega$. On peut ainsi définir un facteur d'automorphie :

$$j_m : \begin{array}{ccc} G \times \Omega & \longrightarrow & \mathbb{C}^* \\ (s,x) & \longmapsto & (cx+d)^m \end{array}$$

pour tout $m \in \mathbb{Z}$.

Si le H-espace Ω est fibré principal de groupe H sur la base $M = H \backslash \Omega$, on peut associer à tout facteur d'automorphie $j : H \times \Omega \rightarrow GL(V)$ un espace fibré différentiable E_j de base M et de fibre V en prenant pour E_j le quotient du fibré trivial $\Omega \times V$ par la relation,, d'équivalence $(x, v) \sim (sx, j(s, x)v)$ associée à l'action de H.

Le diagramme

$$\begin{array}{ccc} \Omega \times V & \longrightarrow & E_j = H\backslash(\Omega \times V) \\ \downarrow \mu_1 & & \downarrow \\ \Omega & \longrightarrow & M \end{array}$$

est commutatif et si l'on abandonne la condition (i) des facteurs d'automorphies on obtient de cette manière tous les fibrés sur M et de fibré V qui sont trivialisés par le changement de base $\Omega \longrightarrow M$.

Si H est un groupe discret Γ tel que $\Gamma \backslash \Omega = M$ soit une variété complexe, le fibré E_j obtenu ci-dessus est un fibré holomorphe sur M. Pour que deux fibrés E_{j_1}, E_{j_2} au-dessus de M soient isomorphes, il faut et il suffit qu'il existe une application holomorphe $f: \Omega \longrightarrow GL(V)$ telle que

(1) $j_2(s, x) = f(sx)^{-1}\, j_1(s, x)\, f(x)$

quels que soient $s \in \Gamma$, $x \in \Omega$.

On dira de manière analogue que deux facteurs d'automorphies $j_1, j_2 : H \times \Omega \longrightarrow GL(V)$ sont <u>équivalents</u> s'il existe une application holomorphe $f : \Omega \longrightarrow GL(V)$ qui vérifie la condition (1) pour $s \in H$

Passons au cas où Ω est un domaine hermitien symétrique muni d'un point base x_0; on peut l'identifier au quotient d'un groupe de Lie semi-simple connexe linéaire G par un sous-groupe compact maximal $K \subset G$, stabilisateur de x_0. De tout facteur d'automorphie $j: G \times \Omega \longrightarrow GL(V)$ on déduit une représentation linéaire de K :

$$\tau : \begin{array}{ccc} K & \longrightarrow & GL(V) \\ t & \longmapsto & j(t, x_0) \end{array}$$

On dira que le facteur d'automorphie j <u>prolonge</u> la représentation τ à Ω. Etant donnés (Ω, x_0) comme ci-dessus, on peut se demander si toute représentation linéaire $\tau: K \longrightarrow GL(V)$ peut être prolongée à Ω en un facteur d'automorphie ; voici la réponse:

<u>Théorème</u>: Soit Ω un domaine hermitien symétrique considéré comme quotient G/K d'un groupe de Lie linéaire semi-simple connexe G par un sous-groupe compact maximal K, muni d'un point base $x_o = K$.

Toute représentation linéaire $\tau : K \longrightarrow GL(V)$ se prolonge de manière canonique en un facteur d'automorphie

$$J_\tau \quad : G \times \Omega \longrightarrow GL(V)$$

On dira que J_τ est un <u>facteur canonique de type</u> τ sur Ω.

Indiquons sommairement la construction de J_τ faite en détail dans [M].

Le plongement de G dans son complexifié $G^{\mathbb{C}}$ se factorise via la partie ouverte $N^+ K^{\mathbb{C}} N^-$ (voir [K]) et posant $U = K^{\mathbb{C}} N^-$, on a $G \subset N^+ U$, $G \cap U = K$ et tout $s \in G$ se décompose de manière unique en $s = n(s).u(s)$ où $n(s) \in N^+$, $u(s) \in U$. Comme $n(su) = n(s)$ pour $u \in K$, l'application $s \longmapsto n(s)$ passe au quotient:

$$\begin{array}{ccc} G & \xrightarrow{\;n\;} & N^+ \\ \downarrow & s \longmapsto n(s) & \nearrow \tilde{n} \\ GU/U \simeq G/K & & \end{array} \qquad \begin{array}{l} \text{si } x_o = K,\ x = sx_o \\ \text{alors } \tilde{n}(x) = n(s). \end{array}$$

Pour $s \in G$ on a $s\tilde{n}(x) \equiv \tilde{n}(sx)$ (mod U) dans $G^{\mathbb{C}}$ et comme $s\tilde{n}(x) \in N^+U = N^+K^{\mathbb{C}}N^-$, il existe un unique $J(s,x) \in K^{\mathbb{C}}$ tel que $s\tilde{n}(x) = \tilde{n}(sx)J(s,x)n'$ ou $n' \in N^-$; cela définit une aplication $J : G \times \Omega \to K^{\mathbb{C}}$ dont on vérifie qu'elle est différentiable et qu'elle satisfait aux conditions :

(i) $\Omega \longrightarrow K^{\mathbb{C}}$ est holomorphe pour
$x \longmapsto J(s,x)$
tout $s \in G$

(ii) $J(ss',x) = J(s,s'x)\,J(s',x)$ quels que soient
$x \in \Omega$ et s, s' dans G

(iii) $J(s,x_o)$ est la composante de s dans $K^{\mathbb{C}}$:
$s = n(s)\,J(s,x_o)\,n'$

et cette dernière relation implique:

(iv) $J(t, x_o) = t$ si $t \in K$.

Comme toute représentation linéaire $\tau : K \longrightarrow GL(V)$ se prolonge en une représentation holomorphe $\tau^{\mathbb{C}} : K^{\mathbb{C}} \longrightarrow GL(V)$, on peut définir le facteur d'automorphie canonique J_τ de type τ par la formule :

$$J_\tau(s, x) = \tau^{\mathbb{C}}(J(s, x)) \text{ quels que soient } s \in G,\ x \in \Omega .$$

Le resultat précédent s'améliore de la manière suivante :

Théorème 2 : Soit Ω un G-espace homogène quotient d'un groupe de Lie connexe G par un sous-groupe fermé K vérifiant les conditions suivantes :

(a) Ω admet une structure complexe invariante par G;

(b) il existe un élément z dans le centre de K tel que la différentielle dL_{z, x_o} soit l'homothétie de rapport -1 dans l'espace tengent $T_{x_o}(\Omega)$. Les facteurs d'automorphie j du G-espace Ω prolongeant une représentation irréductible $\tau : K \longrightarrow GL(V)$ sont tous équivalents.

Donnons les grandes lignes de la démonstration faite dans [M] (appendice) . Soient $j_1, j_2 : G \times \Omega \longrightarrow GL(V)$ deux facteurs d'automorphie prolongeant τ; l'application

$$\tilde{f} \quad \begin{array}{ccc} G & \longrightarrow & GL(V) \\ s & \longmapsto & j_1(s, x_o)\, j_2(s, x_o)^{-1} \end{array}$$

est différentiable et comme $\tilde{f}(st) = \tilde{f}(s)$ pour $t \in K$, elle passe au quotient :

$$\begin{array}{ccc} G & \xrightarrow{\ \tilde{f}\ } & GL(V) \\ \downarrow & \nearrow_{f} & \\ \Omega \simeq G/K & & \end{array}$$

fournissant une application vérifiant (1) . On vérifie ensuite que f est holomorphe en montrant que d" f = o pour l'opérateur d" défini par la structure complexe de Ω . Par différentiation de (1) on fabrique une forme $\phi = f^{-1} d''f$ de type (o, 1) , telle que

$$(\phi \circ dL_s)_x = j_2(s,x)\, \phi_x\, j_2(s,x)^{-1}.$$

Comme j_2 prolonge τ la relation ci-dessus implique en particulier que

$$(\phi \circ dL_t)_{x_o} = \tau(t)\, \phi_{x_o}\, \tau(t)^{-1} \text{ quel que soit } t \in K$$

τ étant irréductible on en déduit pour l'élément z du centre de K :

$$\phi_{x_o} = \tau(z)\, \phi_{x_o}\, \tau(z)^{-1} = (\phi \circ dL_z)_{x_o} = -\phi_{x_o}$$

donc $\phi_{x_o} = o$. La transitivité de G dans Ω entraîne alors que $\phi = o$.

Les hypothèse du théorème 2 sont satisfaites dans la situation du théorème 1, en prenant comme élément z l'élément exp π Z du sous-groupe K, où Z est l'élément de la sous-algèbre $\mathfrak{k}$ qui définit la structure complexe sur G/K .

Corollaire 1: Sous les hypothèses du théorème 1, tout facteur d'automorphie j prolongeant une représentation irréductible de K est équivalent à un facteur canonique J_τ de type τ.

Corollaire 2: Soit A^* le G-module des fonctions holomorphes $\Omega \longrightarrow \mathbb{C}^*$ et m la dimension du centre de K . Le groupe de cohomologie $H^1(G, A^*)$ est isomorphe à $\mathbb{Z}^m$

Cela résulte de ce que $H^1(G, A^*)$ est isomorphe au groupe des caractères de K .

Remarques. 3) Pour le demi-plan de Poincaré les représentations irréductibles de K = SO (2) sont de la forme $\tau : \theta \longrightarrow e^{2\pi i m\theta}$ et un facteur d'automorphie canonique correspondant vaut

$$J_\tau\left(\begin{pmatrix} a & b \\ c & d \end{pmatrix}, z\right) = (cz+d)^m \qquad m \in \mathbb{Z} .$$

4) Dans le theoreme 2 l'hypothese que la representation $\tau : K \longrightarrow GL(V)$ est irreductible est essentielle ; pour un contre-exemple voir [M].

5) Dans le cas des domaines hermitiens irreductibles de type III (espaces de Siegel), les résultats du theoreme 1 avaient été explicites par GUNNING [G].

Bibliographie.

[M] S. MURAKAMI : Cohomology of vector valued forms on symmetric spaces.
notes by F. Grosshans, University of Chicago (summer 1966)

[K] A. KORANYI : Cours du C.I.M.E. 1967

[G] R. C. GUNNING: Homogeneous symplectic multipliers. Illinois J. of Math. 4 (1960) 575-583.

CENTRO INTERNAZIONALE MATEMATICO ESTIVO

(C.I.M.E.)

E.M. STEIN

"THE ANALOGUES OF FATOUS'S THEOREM AND ESTIMATES FOR MAXIMAL FUNCTIONS"

Corso tenuto ad Urbino dal 5 al 13 luglio 1967

THE ANALOGUES OF FATOU'S THEOREM AND ESTIMATES FOR MAXIMAL FUNCTIONS

by

E. M. Stein

(Princeton-University)

1. Introduction.

The behaviour of harmonic functions (in particular Poisson integrals) near the boundary is closely related to the differentiability properties of the boundary funcions. This was long ago recognized in the classical context of the half-plane or disc in Fatou's theorem, and in this setting it was put in more definitive form in terms of the appropriate "maximal functions" first studied by Hardy and Littlewood.

The purpose of this note is to briefly review this, and later developments including those which deal with the product of half-planes or discs, together with the recent attack on the analogous problems for the general symmetric spaces.

While we shall state some new results, we shall not attempt to give proofs here ; those will be published elesewhere.

2. The classical case of the upper half-plane.

The Poisson integrals are those harmonic functions $u(x,y)$ given by

$$(2.1)\qquad u(x,y) = \frac{y}{\pi}\int_{-\infty}^{\infty}\frac{f(x-t)}{y^2+t^2}\,dt = P_y * f\ ,$$

with $y > 0$, $P_y(t) = \frac{b}{\pi}\,\frac{1}{y^2+t^2}$, the convolution $*$ being with respect to additive group of $\mathbb{R}^1$. $u(x,y)$ is harmonic in the ordinary sense. f belongs to an $L^p(-\infty, \infty)$ space, or could be replaced by a finite Borel measure. We shall not dwell here on the rather simple condition that characterizes those harmonic $u(x,y)$ which can be represented in the form (2.1).

The differentiability property of the integral of f, mentioned above is that for $f \in L^p(-\infty,\infty)$, $1 < p \leqslant \infty$

$$\lim_{r \to 0} \frac{1}{2r} \int_{|t| \leqslant r} f(x-t)dt = f(x) , \quad \text{for almost every } x. \tag{2.2}$$

As a simple consequence of (2.2) there is a stronger version of it, namely

$$\lim_{r \to 0} \frac{1}{2r} \int_{|t| \leqslant r} \left| f(x-t) - f(x) \right| dt = 0 , \quad \text{for almost every } x . \tag{2.3}$$

Now as Fatou's theorem showed, whenever (2.2) holds for a given x, then $\lim_{y \to 0} u(x,y) = f(x)$, for x ; also whenever (2.3) holds for a given x, then $\lim u(x',y) = f(x)$, where now (x',y) is allowed to approch $(x,0)$ non-tangentially (i.e. $|x-x'| \leqslant c\, y$, for some constant c).

3. Maximal functions; one dimension.

These matters can be put in sharper relief by the aid of the corresponding "maximal functions". For the process of differentiation of the integral (i.e. (2.2) or (2.3)) the corresponding maximal function is

(3.1) $$f^{*}(x) = \sup_{r > o} \frac{1}{2^r} \int_{|t| \leq r} f(x-t)\, dt$$

(The importance here is that we take $\sup_{r > o}$, instead of $\lim_{r \to o}$ or $\limsup_{r \to o}$).

For this maximal function the following basic facts are well-known.

Theorem A.

If $f \in L^p(-\infty, \infty)$ $1 < p < \infty$, then

(3.2) $$\| f^{*} \|_p \leq A_p \| f \|_p$$

If $f \in L^1(-\infty, \infty)$, then

(3.2') measure $\{ x \mid f^{*}(x) > \alpha \} \leq (A/\alpha) \| f \|_1$.

Here $\| \cdot \|_p$ denotes the usual L^p norm.

The sighificance of this result is as follows: (i) clearly $f^{*}(x) \geq |f(x)|$, but the theorem shows that the reverse inequality also holds, but in a suitably modified functional form. (3.2') is the analogue for $p = 1$ that is weaker than (3.2); (it is also referred to as a "weak-type" inequality). However this form is inherent in the nature of things and cannot be improved. (ii) (3.2) and (3.2') are the quantitative statements whose qualitative analogues are (2.2) and (2.3). More precisely, using theorem A, the fact that differentiability of integrals obviously holds for continuous functions, and some general principles of functional analysis, we can deduce the differentiation theorems form theorem A. Thus we can say, without much oversimplification, that the maximal theorem A contains the essence of the results (2.2) and (2.3).

The connection with Poisson integrals is simple. Here we can also define an appropriate maximal function, that is $\sup_{y>0} |u(x,y)|$, and in fact we have

(3.3) $$\sup_{y>0} |u(x,y)| \leqslant A f^{*}(x)$$

and more generally as far as non-tangential convergence is concerned

(3.4) $$\sup_{|t|\leqslant cy} |u(x-t,y)| \leqslant A_c f^{*}(x), \quad \text{for each } c > 0 .$$

It should be added, as Paley has pointed out, that if $f \geqslant 0$, then the converse implication holds, i.e.

$$f^{*}(x) \leqslant A \sup_{y>0} u(x,y), \quad f \geqslant 0 .$$

Now by the use of (3.3), (3.4), and theorem A we obtain analogous estimates for $\sup_{y>0} |u(x,y)|$, and $\sup_{|t|<cy} |u(x-t,y)|$

Then as in the case of differentiation this leads to the results of almost every-where convergence (even non tangentially) of Poisson integrals contained in Fatou's theorem. So here again the basic results are in fact contained in terms of the appropriate maximal functions. Proofs of all these results may be found in Zygmund's book [8], volume 1.

4. The n-dimensional Euclidean case; a nilpotent variant.

We shall now consider the analogues of these matters but

where the real line is replaced by $\mathbb{R}^n$, or some more general group . We begin with $\mathbb{R}^n$. If $f \in L^p(\mathbb{R}^n)$ we define the analogue of (3.1) by

$$(4.1) \qquad f^*(x) = \sup_{r>0} \frac{1}{m(B_r)} \int_{B_r} |f(x-t)|\,dt$$

where B_r denotes the ball of radius r centered at the origin , and $m(B_r)$ is its Euclidean measure. Then as in the case of n = 1 , the following is well-known.

Theorem B. With f^* defined as in (4.1) the results of theorem A hold for $\mathbb{R}^n$ as well.

This was proved by Wiener [6] , who began by proving the following covering lemma of "Vitali type" .

Lemma. Suppose E is a measurable set of finite measure and $\{B_\alpha\}$ is a collection of balls that cover E. There is a disjoint sub-collection $B_1, B_2, \dots B_n, \dots$, so that

$$\sum_{r=1}^{\infty} m(B_r) \geq c\, m(E) \text{ , where c is an absolute positive constant.}$$

With this lemma (3.2′) follows immediately, and then (3.2) can be deduced from it .

The maximal function allows several variants, which we describe in order of increasing generality. The simplest change, and this is trivial, is to replace the balls appearing in (4.1) by cubes. The next variant is as follows. We choose a fixed rectangle R, and denote by R_r the rectangle obtained by dilating R by the factor r; we consider

$$(4.2) \qquad f^*_R(x) = \sup_{r>0} \frac{1}{m(R_r)} \int_{R_r} |f(x-t)|\,dt \quad .$$

This maximal function again satisfies the conclusion of theorem B, and what is important, with bounds independent of the original rectangle R. This form of the maximal theorem follows from the special case of the cube by a linear change of variables. A further extension is obtained by considering (4.2) again but where $\{R_r\}$ is an arbitrary "monotonic" family of rectangles; i.e. $R_{r_1} \subset R_{r_2}$ if $r_1 \leq r_2$. Here again the bounds do not depend on the family of rectangles. (This form may be found in Zygmund's book [8], chapter 17.). This variant, as well as others by K. Smith [4], and Edwards and Hewitt [2], are proved by adopting the proof of the covering lemma cited above. We shall cite here another generalization, because it is not directly contained in those already mentioned, and is particularly useful in the context of the domains discussed in this conference.

We let G be any locally compact group, with right-invariant Haar measure dm. We assume that there is given a mapping $t \to \alpha_t$ from the positive reals t to automorphism α_t of G, so that $\alpha_{t_1} \circ \alpha_{t_2} = \alpha_{t_1 t_2}$. We assume also that there is a open bounded neighbourhood U of the identity on which the α_t are contracting, i.e. $\alpha_t(U) \subset U$, if $t \leq 1$, and $\bigcup_{t>0} \alpha_t(U) = \{c\}$. (These conditions can be relaxed). Define $f^*(x)$ by

$$f^*(x) = \sup_{t>0} \frac{1}{m(\alpha_t(U))} \int_{\alpha_t(U)} f(y\,x)\,dm(y) \tag{4.3}$$

<u>Theorem 1</u>. <u>The results of theorem</u> A <u>hold equally well in the case of</u> f^* <u>defined by (4.3)</u>.

It goes without saying that each of the maximal theorems discussed in this section implies (by the arguments mentioned in section 3) a cor-

responding result for the existence of limits a.e.
Thus in the case just cited, we have that for each $f \in L^p(G)$ $1 \leq p \leq \infty$,

then $\lim_{t \to o} \frac{1}{m(\alpha_t(U))} \int_{\alpha_t(U)} f(y\,x)\,dm(y) = f(x)$ almost everywhere.

Among the applications of this theorem, which occur typically when G is a nilpotent Lie group, we mention two. In the first example G is still $\mathbb{R}^n$, and if $x = (x_1, x_2, \dots x_n) \in \mathbb{R}^n$, then $\alpha_t(x) = (x_1 t^{a_1}, x_2 t^{a_2}, \dots x_n t^{a_n})$, where the $a_i > o$, but the a_i not necessarily the same. This type of situation occurs when one considers "mixed homogeneity" – as in certain aspect of the theory of singular integrals.

Other applications occur when G is a properly nilpotent group. For example, let G be the group of matricies of the form

$$\begin{Bmatrix} 1 & o & o \\ x_1 & 1 & o \\ x_2 & x_3 & 1 \end{Bmatrix}, \text{ with the automorphisms } \begin{Bmatrix} 1 & o & o \\ x_1 & 1 & o \\ x_2 & x_3 & 1 \end{Bmatrix} \to \begin{Bmatrix} 1 & o & o \\ tx_1 & 1 & o \\ t^2 x_2 & tx_3 & 1 \end{Bmatrix}.$$

The relevance of all of this to Poisson integrals will now be explained.

First of all, theorem B (valid for $\mathbb{R}^n$) is intimately connected with (Euclidean) harmonic functions in the n + 1 dimensional half-space R^{n+1}_+. Thus if we write $(x, y) \in \mathbb{R}^{n+1}_+$, $x \in R^n$, $y > o$, and define $u(x, y)$ by

$$(4.4) \qquad u(x, y) = c_n y \int \frac{f(x-t)}{(y^2 + |t|^2)^{\frac{n+1}{2}}}\,dt, \quad y > o, \quad c_n = \frac{\Gamma(\frac{n+1}{2})}{\pi^{\frac{n+1}{2}}},$$

we say that u is the Poisson integral of f. Here $f \in L^p(\mathbb{R}^n)$, and u is harmonic in the sense that it is annihilated by the Laplacean

$$\frac{\partial}{\partial y^2} + \frac{\partial}{\partial x_1^2} + \dots + \frac{\partial}{\partial x_n^2}$$

. Now as in the case of one variable

(4.5) $$\sup_{y>0} |u(x,y)| \leq A\, f^*(x) ,$$

where f^* is given by (4.1).

There is a similar inequality for non-tangential behaviour as well as the fact that $f^*(x) \leq A \sup_{y>0} u(x,y)$, if $f > 0$.

Because of (4.5), and the maximal theorem, theorem B, one obtains the convergence a.e. of Poisson integrals of functions in $L^p(\mathbb{R}^n)$ $1 \leq p \leq \infty$. Theorem 1 plays a similar, although less decisive role, for Poisson integrals in the case of non-compact symmetric spaces and generalized half-planes (i.e. "Siegel domains of type II"). In the first case the reason for this is indicated by the Iwasawa decomposition $G = KAN$, and the fact that the boundary (the so called Furstenberg boundary) can be essentially identified with the nilpotent group N. In the case of the generalized half-planes the distinghuished boundary can also be identified with a nilpotent group. The convergence theorem for generalized half-planes is stated in Koranyi's lectures given at this conference, and the proof will be published in a joint paper of Koranyi and the author. It must be pointed out, however, that in both cases the results are only for the special case of Poisson integrals of bounded founctions.

What is probably needed for the general case is a more refined version of theorem 1. A hint of the ultimate version for the general case of nilpotent groups is given by the refinements for $\mathbb{R}^n$ we shall now discuss.

5. The case of product of half-planes, and some other domains.

The different maximal functions treated so far have essentially the same real-variable character whether they be for $\mathbb{R}^1$ in section 3, $\mathbb{R}^n$ or G in section 4. These matters begin to change, however, as soon as we consider the simplest product domains.

Since in the case $n = 1$, the differentiation of integrals is carried out with respect to intervals, so in the product case we differentiate with respect to their products, i.e. "rectangles". That is, we still consider $\mathbb{R}^n$, and $f \in L^p(\mathbb{R}^n)$, and now pose the question whether

(5.1) $$\lim_{R \to o} \frac{1}{m(R)} \int_R f(x-t)dt = f(x), \quad \text{almost everywhere,}$$

where R runs over the family of rectangles with sides parallel to the axes which contain the origin; $R \to o$ means the diameter of R tends to zero. To study this question we consider the corresponding maximal function

(5.2) $$f^{**}(x) = \sup_R \frac{1}{m(R)} \int_R |f(x-t)| dt .$$

Let us state right away that the results here are different from what we had up to this point. In fact there exists an $f \in L^1(\mathbb{R}^n)$, $(n > 1)$, so that $f^{**}(x) = \infty$ everywhere, and in particular, the limit (5.1) exists nowhere. In this way we see that this type of differentiation, involving rectangles, is more hazardous that the one involving cubes or balls which was discussed in section 4. We refer to the present type as <u>strong differentiation</u> and the previous type as <u>ordinary differentiation.</u> However positive results do hold for $L^p(\mathbb{R}^n)$, $p > 1$, for strong differentiation.

Theorem G

(a) $\|f^{**}\|_p \leqslant A_p \|f\|_p$, $1 < p < \infty$.

(b) If $f \in L^p(\mathbb{R}^n)$ $p > 1$, then (5.1) holds.

This theorem is due to Zygmund. One way of proving it is by observing that the rectangular function f^{**} is obtained by superposition of one-dimensional maximal functions n times, (once for each dimension). Since each one-dimensional maximal function preserves the class L^p , $p > 1$, the process of superposition can be carried out. However if $f \in L^1$, then (3.2') takes us out of the class of integrable functions, and so the process ends at the end of the first step.

Actually, the real dividing line between L^p and L^1, as far as the conclusion (5.1) is concerned is the class $L(\log L)^{n-1}$ but we shall not discuss this point further.

These results are related to harmonic functions on the product of half-planes as follows. The product of n half-planes is the tube domain $T_\Gamma = \{z = x+iy,\ x \in \mathbb{R}^n,\ y \in \Gamma$ where Γ is the cone which is the first "octant", i.e. $\Gamma = \{y,\ y_j > 0,\ j = 1, \dots n$.

If $f \in L^p(\mathbb{R}^n)$ its Poisson integral with respect to this domain is

(5.3) $$u(x,y) = \int_{\mathbb{R}^n} P_y^\Gamma(t)\, f(x-t)\, dt$$

$P_y^\Gamma(t)$ is the product of one-dimensional Poisson kernels

$$P_y^\Gamma(t) = \pi^{-n} \prod_{j=1}^{n} \left(\frac{y_j}{y_j^2 + t_j^2} \right),$$

As in the case of one variable, it is an elementary fact that

(5.4) $$\sup_{y \in \Gamma} |u(x,y)| \leqslant A\, f^{**}(x) ,$$

with a similar result for non-tangential approach. Also if $f \geq o$, then

$$f^{**}(x) \leq A \sup_{y \in \Gamma} u(x,y) .$$

From theorem G, (5.4), and the arguments already collected the following corollary of theorem G may be proved.

Corollary.

If $f \in L^p(\mathbb{R}^n)$, $p > 1$, then,

$$\lim_{\substack{y \in \Gamma \\ y \to o}} u(x,y) = f(x) \quad \text{almost everywhere.} \tag{5.5}$$

As in the case of strong differentiation, there are examples which show that the conclusion of the corollary fails for the class $L^1(\mathbb{R}^n)$. If one compares the notion of strong differentiation to ordinary differentiation one is tempted to modify the limit in (5.5) by requiring that $y \to o$, in such a way that the components y_j of y are of equal order of magnitude. That is, by restricting y to lie in a proper subcone Γ_o of Γ, (i.e. $\Gamma_o \subset \Gamma \cup$ origin). This is the setting for the theorem of Marcinkiewicz and Zygmund.

Theorem D. Let Γ be the first octant, thus T_Γ = product of half-planes. Let Γ_o be a proper sub-cone of Γ. Then if $f \in L^p(\mathbb{R}^n)$ $1 < p < \infty$.

$$\lim_{\substack{y \in \Gamma_o \\ y \to o}} u(x,y) = f(x), \quad \text{almost everywhere.} \tag{5.6}$$

For obvious reasons we refer to the convergence (5.5) as unrestricted convergence and (5.6) as restricted convergence. (Up to this point all the results of this section may be found in Zygmund's book, chapter 17).

We come now to the general situation. While analogues of such results might be presumed to hold in a rather wide context, i.e. including all non-compact symmetric spaces, and generalized half-planes (Siegel domains of type II) we shall content ourselves with the description of a stage of intermediate generality. This situation is probably already indicative of the general context, and in addition can be described without going too far afield. Thus we shall limit ourselves to tube domains T_{Γ} , which are domains of positivity, i.e. those tube domains whose basis Γ is a homogeneous self dual cone. These tube-domains represent an important sub-Class of the bounded symmetric domains.

Thus if Γ is such a cone, R^n, and $f \in L^P(\mathbb{R}^n)$ we form the Poisson integral of f with respect to the cone

$$u_{\Gamma}(x, y) = \int_{\mathbb{R}^n} P_y^{\Gamma}(t) \, f(x-t) dt ,$$

where $P_y^{\Gamma}(t)$ is the Poisson kernel of the tube domain T_{Γ} (Stein, G. Weiss, and M. Weiss [5] , and Koranyi [3]). u (x, y) is then harmonic with respect to the invariant operators of the bounded symmetric domain T_{Γ} .

We may then pose the problems of unrestricted and restricted convergence as in (5.5) and (5;6) .

The first fact is that unrestricted convergence does not seem to be appropriate in this general context because as is shown in [5] , for every $p < \infty$ there exists a tube domain of this type, T_{Γ} , so that unrestricted almost everywhere convergence fails for L^p . However for restricted convergence the results are more positive. As N. Weiss showed in his thesis [7], restricted convergence holds for these tube domains for L^p, $p > 1$; and even for the class $L(\log L)$.

The problem that remains is, therefore, the problem for L^1. A clearer understanding of this problem, leading to a positive solution, will be discussed in the next section.

6. The L^1 case.

We begin by going back to the special case when Γ in the octant $y_j > o$, s $= 1, \dots n$; i.e. T_Γ = product of half-pplanes. Let Γ_o be a proper sub-cone of Γ. Then if $f \in L^1(\mathbb{R}^n)$ and $u(x,y)$ is its Poisson integral, then as is shown in the proof of theorem D, $\sup_{y \in \Gamma_o} |u(x,y)|$ satisfies an estimate analogous to (3.2'), i.e.

(6.1) $$m\left\{\sup_{y \in \Gamma_o} |u(x,y)| > \alpha\right\} \leq A/\alpha \|f\|_1 .$$

It would therefore be of interest to deduce this fact as a consequence of the same kind of estimate for a maximal function of the type (4.1), or some variant of it. Indeed, as is easy to see if $f \geqslant o$, then with f^* defined as in (4.1) $f^*(x) \leq A \sup_{y \in \Gamma_o} u(x,y)$, but it is definitely not true that $\sup_{y \in \Gamma_o} |u(x,y)| \leq A f^*(x)$. We shall now describe how f^* must be modified.

It is interesting to point out that the right variant (4.1) for this type of problem has been introduced some time ago in another connection, in analogy with certain singular integrals. In fact let $\Omega(x)$ be a non-negative function homogeneous of degree zero in $\mathbb{R}^n$, i.e. $\Omega(x) = \Omega(\lambda x)$, $\lambda > o$. which is therefore determined by its values on the unit sphere. Assume that Ω is integrable on the unit sphere. In terms of Ω consider the maximal function

(6.2) $$f^{*}_{\Omega}(x) = \sup_{r>0} \frac{1}{r^n} \int_{|t| \leq r} |f(x-t)| \, \Omega(t) \, dt .$$

When Ω is constant, then we get to (4.1). For general Ω we are dealing with a maximal function that has certain preferential directions, those where Ω is relatively large. Now the following fact was proved about f^{*}_{Ω} by Calderon and Zygmund [1].

<u>Theorem E. Suppose</u> $f \in L^p(\mathbb{R}^n)$, $1 < p < \infty$, <u>then</u>

(6.3) $$\| f^{*}_{\Omega} \|_p \leq A_p \| f \|_p .$$

The idea of the proof of this theorem is to use the one-dimensional L^p result (3.2) in each direction emenating from the origin, and then integrate over all these directions. However an L^1 result cannot be obtained in this way, because the L^1 inequality (3.2'). involves the notion of weak-type 1, which in distinction to the case of a norm is not sub-additive. We shall return to this point nomentarily, but we deal first with the relation between $\sup_{y \in \Gamma_0} u(x,y)$ and $f^{*}_{\Omega}(x)$.

<u>Theorem 2. Suppose Γ is a classical homogeneous self-dual cone, $\Gamma \subset \mathbb{R}^n$. Then there exists a positive function Ω, homogeneous of degree 0, and integrable over the unit sphere, so that if Γ_0 is a proper sub-cone of Γ, then</u>

$$\sup_{y \in \Gamma_0} | u(x,y) | \leq A \, f^{*}_{\Omega}(x) .$$

The word "classical" indicates that this theorem has been verified (by direct computation) in all but the exceptional case. But there can be no doubt that it holds also in this case.

We give here two examples

Example 1. Γ = octant, $y_j > o$, $j = 1, \ldots n$, in $\mathbb{R}^n$.

Then we can take $\Omega(x) = \left(\frac{|x_1 \cdot x_2 \cdot \ldots, x_n|}{|x|^n} \right)^{-\varepsilon}$, where

$|x| = (x_1^2 + x_2^2 \ldots + x_n^2)^{1/2}$, $\varepsilon > o$.

Example 2. Γ = n-dimensional circular cone, i.e.

$$\Gamma = \left\{ (y_1, \ldots, y_n), \text{ where } y_1^2 > y_2^2 + \ldots + y_n^2, \; y_1 > o \right\}$$

Then we can take $\Omega(x) = \left| \frac{|x_1|}{|x|} - \frac{\sqrt{2}}{2} \right|^{-\varepsilon}$, $\varepsilon > o$.

We see therefore that the L^1 problem for Poisson integrals may be reduced to a similar L^1 problem, but for the maximal function $f^*_\Omega(x)$, (6.2).

While we do not solve the L^1 problem for f^*_Ω for general Ω, we can solve it for a wide variety of Ω which include all those arising in thorem 2.

The main technique is the following.

Suppose $\left\{ f_j(x) \right\}_{j=1}^{\infty}$ is a sequence of positive functions, uniformly of weak-type 1, i.e. $m \left\{ f_j(x) > \alpha \right\} \leq 1/$, for each j. Suppose a_j are positive constants ; what can be said of

$$f(x) = \sum_{j=1}^{\infty} a_j f_j(x) \quad ? \tag{6.4}$$

If the class of functions which are of weak-type 1 could be normed (and this is not the case), then the condition $\sum a_j < \infty$ would suffice to imply that f is itself of weak type 1. However a substitute result holds.

Lemma 1. Suppose $q < 1$, and $\sum_{j=1}^{\infty} (a_j)^q \leq 1$.

Then $m\{f(x) > \alpha\} \leq A_q/\alpha, \ \alpha > 0$.

Whith the aid of this lemma we can prove

Theorem 3. Under the assumptions of theorem 2,

$$\lim_{\substack{y \in \Gamma_0 \\ y \to 0}} u(x, y) = f(x) \quad \text{holds almost everywhere,}$$

if $f \in L^p(\mathbb{R}^n)$, $1 < p < \infty$

The proof of lemma 1 as well as theorems 1, 2, and 3 will be given in detail elsewhere.

REFERENCES

1 A. CALDERON and A. ZYGMUND, On singular integrals, Amer. Math. , 78 (1956) , 289-309 .

2 R. E. EDWARDS and E. HEWITT, Point wise limits for sequences of convolution operators, Acta Mathematica, 113 (1965) 181-218

3 A KORANYI, The Poisson integral for generalized half-planes and bounded symmetric domains, Annals of Math. 82 (1965) pp. 332-350.

4 K. T. SMITH, A generalization of an inequality of Hardy and littlewood, Canad. J. Math. 8 (1956) , 157-170 .

5 E. M. STEIN , G. WEISS, and M. WEISS, H^p classes of holomorphic functions in tube domains, Proc. Nat. Acad. Sci . USA, 52 (1964) , 1035-1039 .

6 N. WIENER, The ergodic theorem, Duke Math. J. 5(1939) , 1-18 ; Selected papers of Norbert Wiener, The M. I. T. Press. 1964, 412-429.

7 N. WEISS, Doctoral dissertation 1966, to appear Trans. Amer. Math Soc.

8 A. ZYGMUND , 'Trigonometric series' , Cambridge, 1959.

STAMPA EDITORIALE GRAFICA - ROMA